IBDP STUDY GUIDE
CHINESE B
WRITING
Simplified Characters

For **NEW** syllabus

 NTK Publishing Limited

IBDP STUDY GUIDE
CHINESE B
WRITING
Simplified Characters

For **NEW** syllabus

 NTK Publishing Limited

The Chinese Department at NTK Learning Center consists of a team of experienced and professional subject specialists who are passionate about teaching Chinese language and culture. With a wealth of experience in teaching IB Chinese, NTK has helped hundreds of students achieve outstanding IB exam results. The authors and contributors of this study guide are currently teachers in our Chinese Department:

Authors

Ling Tsz-wai (BA, MA, PhD)

Hu Xiao-qian (BA, MA)

Han Ying (BA, MA)

Contributors

Zhang Rui (BA, MA)   Jiang Fan (BA, MA)

Zhong Zhi-min (BA, MA)   Wang Ning-ning (BA, MA)

Liu Jia (BA, MA)   Wu Tian-ning (BA, MA)

First published 2014
Reprinted 2016
©NTK Publishing Limited (Part of NTK Academic Group)

Disclaimer:
Information is current as of publication date. Every effort has been made to publish this book as complete and accurate as possible. The information provided is on an "as is" basis. The authors and publisher shall have no liability or responsibility for any loss or damages arising from the contents of this publication.

All figures used in this book are used with the express agreement from independent designers that the artwork is original. The publisher is indemnified from any copyright issues related to any such artwork.

The material in this publication has been developed independently by the publisher and the content is in no way connected with nor endorsed by the International Baccalaureate Organization.

Published by:
NTK Publishing Limited
7/F, 18 Hysan Avenue,
Causeway Bay,
Hong Kong SAR

Tel: +852 2577 7844
Fax: +852 2881 6708
E-mail: publishing@ntk.edu.hk
ISBN 978-988-15555-4-0
Website: www.ntk.edu.hk

 www.facebook.com/NTKAcademicGroup

 www.youtube.com/user/NTKAcademicGroup

■ Foreword

Academic success can be measured in many different ways, and I often tell my students that scoring high marks in exams is only one of the rewards of diligent study. The true measures of academic success should be the enjoyment of learning and the sense of accomplishment students get when that light flicks on in their heads and they think to themselves, "So that's why!" The inception of NTK's study guides and publications is based on the simple goal of making the learning process more enjoyable and less complicated, while delivering positive results from students' efforts.

A language, whether foreign or native, cannot be mastered in a single day. Focus and persistence are the keys to mastering a language, and no amount of studying can replace a good learning attitude and regular use of language. NTK's study guides, courses and educational services are designed to help prepare students for exam success as they continue to pursue secondary and college education.

As a leading educational service provider in Southeast Asia since 1996, NTK has helped thousands of students reach their academic goals. Whether they are in primary, secondary, or post-graduate studies, our students have benefited greatly from our specialized academic programs and expertise in all major international curricula and exams.

As you continue on your studies, I wish you every success and most importantly, I hope you enjoy the learning process as well.

T. K. Ng

Founder and Managing Director
NTK Academic Group

■ 序

　　每个人对学业成功有不同的定义，我经常告诉学生考取高分并不是笃学奋进的唯一目标。学业成功的真谛在于：一是求学过程中孜孜不倦、乐此不疲的那种精神；二是突破瓶颈后茅塞顿开的那份成就感。德勤教育集团出版各类教育丛书的初衷，便是鼓励学生乐学好学，使其在驾轻就熟中取得斐然的成绩。

　　无论是学习母语还是外语，都不可能一蹴而就。只有拥有明确清晰的目标，持之以恒的决心，锲而不舍的态度，日积月累的方法，你才能出口成章，妙笔生花。我们的学习指南、各种课程及教学服务便旨在帮助同学们备考应试，最终金榜题名。

　　作为东南亚地区赫赫有名的教学服务集团，我们从1996年起便帮助了数以千计的学生梦想成真。无论是中小学生还是大学生，都从我们为其量身订制的课程中获益匪浅。我们的学生不仅在学校名列前茅，更在各项考试中捷报频传。

　　最后，谨祝同学们学业有成，乐学致远！

黄德勤
德勤教育集团
创办人兼执行总监

■ Foreword 2

Today, IBDP is the world's most influential curriculum and IBDP Chinese B has become a popular subject choice for students. However, the study guides currently available on the market are either overly difficult or too general. To this end, NTK's Chinese Department has comprised its years of teaching experience into producing the IBDP Chinese B Writing Guide, with a specific focus on the exam.

First, 18 practical writing text types are introduced in this study guide. Presented in both Chinese and English, Part 1 introduces commonly-tested genres, including their features, format, tones, etc. With the perfect-scoring sample essays provided, students can learn about different genres and how to establish a strong writing structure.

Second, this study guide provides useful vocabulary and sample sentences based on the 5 option topics. In Part 2, you will find 25 SL/HL sample essays, as well as a bilingual vocabulary list, common grammar and sentence structures.

Third, the trend in popular exam topics is thoroughly analyzed. Part 3 focuses specifically on the 3 core topics, providing useful phrases and stimulation for students to think in multiple perspectives, while strengthening their Chinese expression skills, as well as reading comprehension and oral skills.

Fourth, nearly 150 writing questions are provided for practice. Good writing comes from persistent practice and of course, practice makes perfect.

Fifth, marking criteria and scoring details are summarized at the end. Part 4 is a collection of students' essays with teachers' grading and comments. Students can better understand the assessment criteria and avoid making the same mistakes, enabling them to achieve the highest score.

Chinese language cannot be mastered in a single day, but using proper learning materials can ensure that students are on the right track to success. We sincerely hope that this study guide will help you achieve your best performance in the exam.

Dr. Ling Tsz-wai

Vice Principal (Chinese)

■ 序二

如今，IBDP是全球最具影响力的课程，IBDP Chinese B也成为学生们争相报考的热门科目。但目前市面上的参考书，不是内容艰深，就是泛泛而谈。为此，NTK中文部推出IBDP Chinese B Writing Guide一书，把多年累积下来的教学经验，浓缩提炼成这本极具针对性的写作指南。

第一，详细介绍超过18种实用文体的写作方法。指南的第一部分以中英双语形式介绍常考文类的特色、格式、语气等，并附有满分范文，使学生在掌握不同文体的同时，建立严谨结构。

第二，提供5个"选修主题"相关的常用词句及范例。指南的第二部分收录了25篇SL/HL的写作范例，并附有中英对照的词汇、常用语法及句式。

第三，剖析热门考题的解题思路。指南的第三部分针对3个核心主题，提供大量思考点及实用词组，不仅能激发学生多角度思考，强化他们的中文表意技巧，同时有助于巩固他们中文阅读理解及口试的基础。

第四，提供接近150道的写作训练题目，只要多写多练、持之以恒，必然熟能生巧。

第五，总结评分准则及得分细节。指南的第四部分收录了多篇学生作文，由专业老师评分及解析，让学生们了解各个评分标准，避免重蹈复辙，争取最佳分数。

要学好中文，难以一步登天；但善用工具，可以事半功倍。衷心祝愿本书助您驰骋考场！

凌子威博士

德勤教育集团
副校长(中文)

目录 Contents

Introduction

简 介

简介

本书由德勤教育集团的资深中文教学团队精心编撰，是一本专门针对 IBDP Chinese B 写作部分（试卷2）的应试教材。全书完全根据 IBDP 考试大纲而写，是一本高效实用的应试写作辅导书。

一 本书构成

本书正文分为考试说明、文类解析以及主题范文三部分。

考试说明	详细介绍IBDP Chinese B Standard Level（以下简称SL）和Higher Level（以下简称HL）写作部分的考试范围、答题要求及评分细则。
文类解析	根据最新考纲要求，结合历年考题，为学生总结整理了各种必考文类。
主题范文	严格根据考纲要求，从8个主题：交流与媒体、全球性问题、社会关系、文化的多样性、风俗与传统、健康、休闲、科学与技术着手，并在大主题下对话题进行细分，为学生提供热门试题的满分范文，并提供该范文的词组让学生参考。

二 本书特点

- **紧扣考纲要求**：本书严格按照最新考纲编写，把握最前沿命题方向，并由专业IB导师对其内容进行审定。

- **详细的文类分析**：文类在IBDP Chinese B的考试中占据非常重要的地位。本书从格式、语言、常用句子等多方面对文类进行分析，力求让学生清楚掌握各种文类的写作技巧。每一种文类均附有一篇范文，理论与实践相结合，使学生对知识的掌握更清晰明了。

- **范文全面覆盖主题**：本书所选范文源于考纲规定的3个核心主题和5个选修主题。考虑到SL和HL对语言及内容的要求上存在差异，针对同一主题，本教材分别选取符合各自难易程度的满分范文，并提供实用词组及常用词汇供学生参考。

我们衷心希望这本书可以成为学生的良师益友，帮助学生克服中文写作的难题，轻松应对中文写作，在IBDP Chinese B的考试中取得优异的成绩。

Introduction

This study guide is written by a team of professional Chinese specialists who have many years of experience in teaching IB Chinese. It focuses on the writing assessment (Paper 2) of the IBDP Chinese B exam and is written in accordance with the IBDP syllabus. The book aims to serve as an efficient and practical exam-oriented writing guide.

1   Structure

This writing guide is divided into three sections: exam format, text type analysis and topic-based writing sample.

Exam format	A detailed introduction to the exam scope, requirements and marking scheme for the writing assessment in both the IBDP Chinese B Standard Level (SL) and Higher Level (HL) exams.
Text type analysis	A comprehensive analysis of text types that frequently appear in exams, according to the latest syllabus and questions from past exam papers.
Topic-based writing sample	The writing samples cover 8 major themes, including communication and media, global issues, social relationships, cultural diversity, customs and traditions, health, leisure, and science and technology. High scoring essays and useful phrases will be provided.

2   Features

- **Close alignment to the IBDP syllabus:** this book is written by professional IB experts in accordance with the latest syllabus and exam trends.
- **Detailed text type analysis:** knowing different text types is highly important in the IBDP Chinese B exam. This book introduces various text types to students and analyzes them in terms of format, language, commonly-used sentences, etc. A writing sample is provided for each text type, along with teaching points to enhance students' understanding.
- **Full coverage of exam topics in the writing samples:** the writing samples in this book are based on the three core and five option topics in the syllabus. Catering to the various requirements of language and content between SL and HL, this writing guide provides perfect-scoring samples in two different levels of difficulty for each topic. Useful phrases and expressions are also provided for easy reference.

We sincerely hope that this writing guide will help students overcome the difficulties in writing, tackle written tasks with ease and achieve excellent results in the IBDP Chinese B exam.

IBDP Chinese B写作部分介绍

IBDP Chinese B写作部分（试卷2）占考试总分的25%，考试时间为90分钟。对考生来说，这一部分通常是最困难的，也最能拉开考生之间差距。现根据IBO最新发布的考纲，对IBDP Chinese B的写作部分做一个详尽的介绍。由于SL和HL的考试内容有较大差异，故在此分述其特点。

一 SL

该部分基于IBO所提供的5个选修主题：文化的多样性、风俗与传统、健康、休闲、科学与技术。试卷中有5道写作题，每一道题均基于一个主题，要求学生采用一种具体的文类进行写作，比如书信、演讲稿或者新闻报导等。学生可以从5道题目中选择一道作答。文章篇幅为300-480个汉字之间。

根据评分标准，写作部分将从以下三个方面进行评分，总分为25分。

标准A	语言	10分
标准B	讯息	10分
标准C	形式	5分
	总计	25分

二 HL

HL的写作部分由A和B两个部分组成：

- A部分和SL的作文要求基本相同，须根据IBO所提供的5个选修主题，写一篇300-480个汉字的文章。HL与SL的不同之处主要有两点：第一，HL中可能出现的文类比SL多；第二，HL对学生语言运用及词汇量的要求更高。A部分的评分标准和SL作文的评分标准基本一致。

- 在B部分，学生需要基于与核心主题相关的启发材料做出推理论证性质的应答。根据考纲，IB Chinese B共有3个核心主题：交流与媒体、全球性问题、社会关系。启发材料可能是一篇新闻报导或是一位公众人物发表的见解。该部分不仅考察学生对材料的理解能力，还考察学生的写作能力。学生作答该部分时应当注意文章的细节，并结合自己对该核心主题的理解，表达自己对材料的反思并展开讨论。B部分并没有固定答案，主要考察学生对核心主题的理解以及展开论证的能力。根据评分标准，B部分将从以下两个方面进行评分，总分为20分。

标准A	语言	10分
标准B	论证	10分
	总计	20分

总体说来，IB Chinese B写作部分的关键在于语言和格式，即使用正确的格式、丰富的词汇清晰地表达自己的观点。本书的设计正是从这两点出发，从文类到常用词汇为广大学生做一总结，以期可以帮助学生写出高分作文。

Overview of IBDP Chinese B Writing

The IBDP Chinese B Writing (Paper 2) accounts for 25% of the overall score and is 90 minutes long. This paper is seen as the most difficult for candidates, and it is the best in differentiating candidates according to their performance. The following is an overview of the Writing assessment of the IBDP Chinese B (SL and HL) exam based on the latest syllabus.

1   SL

The SL paper is based on five option topics set by the IBO: cultural diversity, customs and traditions, health, leisure, and science and technology. There are five tasks, and each one is based on a different topic and requires the student to write in a specific text type, such as letter, speech notes, news report, etc. The student has to choose one of the five tasks and write 300-480 words.

The writing paper is assessed according to the following criteria, with 25 marks in total.

Criterion A	Language	10 marks
Criterion B	Content	10 marks
Criterion C	Format	5 marks
	Total	25 marks

2   HL

The HL writing paper consists of two sections:

- Section A is assessed in the same way as that in SL. It is based on the five option topics set by the IBO and requires the candidate to write an essay of 300-480 words. The differences between HL and SL: firstly, HL covers more text types; secondly, HL requires more advanced use of language and vocabulary. Section A has the same assessment criteria as that in SL.

- In section B, the student is required to write a reasoned argument in the form of a response to given prompts related to one of the core topics. The IB Chinese B syllabus has three core topics: communication and media, global issues and social relationships. The prompts could be a news report or a comment by a public figure. This section not only tests students' comprehension of the materials but also assesses their writing ability. In answering this section, students should pay attention to the details, try to understand the core topic in the given prompts, and express their reflection on the text with further discussion. There is no model answer for section B. It mainly assesses the student's ability in understanding and discussing the core topic. Section B is assessed according to the two criteria below, with 20 marks in total.

Criterion A	Language	10 marks
Criterion B	Argument	10 marks
	Total	20 marks

In summary, the IBDP Chinese B Writing assessment weighs heavily on format and language – the student is expected to use the right format and a variety of vocabulary to express their opinions in a clear manner. This book is essentially written with the focus on both language and format. By covering all the text types and providing useful vocabulary, we hope the student will acquire the skills necessary to produce high-scoring essays.

Writing formats and sample essays

文 体 格 式 及 范 文

1 书信

文体介绍

书信就是我们通常说的私人信件，结构如下：

称谓语	要顶格写，后面跟冒号 ":"
	亲爱的爸爸：
	尊敬的校长：
开首语	要空两格写
	你好！好久不见了！你过得好吗？
	你好／你们好！好久不见了！我非常想念你，最近一切都好吗？
正文	要空两格开始写
结束语	要空两格开始写
	好了，不多说了，我要做功课去了！期待你的回信。
	希望我的意见对你有用，等着／期待你的回信。
祝福语	要空格写
	祝你生活愉快／身体健康／学业进步／早日康复／工作顺利！
署名和日期	在右下角写
	正式信件需写全名，非正式信件不冠上姓。
	日期应为年、月、日。

1 Letter

Text type introduction

A general letter is also called a personal letter. It has the following structure:

Salutation	Start at the beginning of the line
	Dear Dad,
	Honorable headmaster,
Opening	Indent two spaces
	Hello! Long time no see! How's life going?
	Hello! Long time no see! I miss you very much. How has everything been recently?
Body	Indent two spaces for each new paragraph
Ending	Indent two spaces
	I still have so much to say, but I have to do my homework now! Looking forward to your reply.
	I hope my advice is helpful to you. Looking forward to your reply.
Blessing	Indent two spaces
	Wish you a cheerful life! / Wish you good health! / Wish you progress in your study! / Get well soon! / Wish everything goes well at work!
Signature and date	Bottom right at the end of the letter
	Write your name in full if it is a formal letter and only your first name if it is an informal letter. Date: Year / month / date

格式 Format

尊敬的校长 / 老师：

亲爱的爸爸 / 老师 / 表姐 / 小明：

1、您好 / 你好 / 你们好！好久不见了！我非常想念你，最近一切都好吗？

2、您好 / 你好 / 你们好！很久没有收到你的来信了，很是挂念。

3、您好 / 你好 / 你们好！你的信我上个星期就收到了，但是我最近很忙，到现在才有空回信，真对不起！你在信中提到……

4、你好 / 你们好！很高兴收到你的信。你在信中提到……

1、好了，不多说了，我要写作业了！期待你的回信。

2、希望我的意见对你有用，等着 / 期待你的回信。

3、对于这件事你有什么看法呢？我很希望听听你的意见。

4、祝你们能够早日恢复正常生活。

　祝

生活愉快，学业进步！/
身体健康，工作顺利！/
早日康复！

儿子 / 朋友 / 学生

名字

X年X月X日

范文 Sample essay

你的朋友小美刚刚失恋了，请写一封信安慰她，并鼓励她好好生活。（SL）

亲爱的小美： — "Dear Xiaomei (小美),"

　　你好！你的来信我上个星期已经收到了，可是我最近很忙，一直没时间回信。真对不起！听说你刚刚失恋，感到很难过，我很担心你。

Greet the letter recipient and state the purpose of the letter.

　　失恋真的会使人很不开心，但是，我们应勇敢地面对，这样才不会让身边的人担心。你现在只是离开了一个不再爱你的人，但你还有爸爸妈妈、老师和很多朋友。他们都很爱你，所以，你拥有的爱比你失去的多。此外，爱情只是我们生活的一部分，除了爱情，我们还有亲情、友情、健康，也有学业和爱好。你是一个好女孩，你将来一定会找到一个更合适你的人。还有，你现在还是学生，很快就要考大学了，应该努力读书，这样才可以实现你当医生的梦想。

Offer words of consolation to Xiaomei (小美).

　　如果你想开心一点，可以多跟父母和朋友谈谈。他们会跟你分享他们的经验和看法，这样你会感觉好一点。你也可以多参加课外活动或去当义工，认识更多不同的人，你会发现生活是美好的。

Give suggestions and advice to Xiaomei (小美).

　　希望我的建议可以帮到你吧！期待你的回信！

Conclude by saying that you look forward to receiving Xiaomei's (小美) reply.

　　祝
生活愉快！ — *Blessing*

朋友
小明 — *Signature and date*
二零一四年六月三日

(348 words)

2 电子邮件

文体介绍

电子邮件指使用者之间通过互联网发出或收到的信件。

电子邮件必须标明以下几个方面：

- 寄件人电邮
- 收件人电邮
- 主题
- 日期（年、月、日 / 星期 / 具体时间）
- 祝词
- 署名

电子邮件的正文和一般的书信格式一样。

2 E-mail

Text type introduction

An e-mail is a letter exchanged between users via the Internet.

In writing an e-mail, the following information should be included:

- Sender's e-mail address
- Recipient's e-mail address
- Subject
- Date (Year - month - date / day / time)
- Blessing
- Signature

The format of an e-mail is the same as that of a letter.

格式 Format

寄件人：xxx@email.com
收件人：yyy@email.com
主题：XXXXXXXX
日期：X年X月X日　星期X　上午 X：X

尊敬的校长 / 老师：
亲爱的爸爸 / 老师 / 表姐 / 小明：

1、您好 / 你好 / 你们好！好久不见了！我非常想念你，最近一切都好吗？
2、您好 / 你好 / 你们好！很久没有收到你的邮件了，很是挂念。
3、您好 / 你好 / 你们好！你的邮件我上个星期就收到了，但是我最近很忙，到现在
　　才有空回邮件，真对不起！你在邮件提到……
4、你好 / 你们好！很高兴收到你的邮件。你在邮件中提到……

1、好了，不多说了，我要写作业了！期待你的回复。
2、希望我的意见对你有用，等着 / 期待你的回复。
3、对于这件事你有什么看法呢？我很希望听听你的意见。
4、祝你们能够早日恢复正常生活。

　祝
生活愉快，学业进步！/
身体健康，工作顺利！/
早日康复！

儿子 / 朋友 / 学生

名字

范文 Sample essay

你是一个关心环保的人。最近，你想买一辆车。写一封电子邮件跟你的网友讨论自己开车的好处及坏处。（SL）

Indentation is not needed.

寄件人：Mary_mary@email.com
收件人：David_david@email.com
主题：开车的问题
日期：2013年12月10日 星期二 下午 12:20

Greet the receiver and express your concern for the environment.

大卫：

　　好久不见了，你好吗？我最近想买一辆新车，可是我很担心会对环境不好，我想听听你的意见。

List the benefits of having your own car.

　　我觉得有自己的车比较方便。因为，上学不用花时间等车，这样我每天可以多睡半个小时。放假的时候，我喜欢去海滩玩。如果没有自己的车，我就要先乘地铁，再搭公共汽车才能到海滩，非常不方便。而且，从我家到海滩的车很少，人又多，我常常要等很久才可以上车。如果我自己开车，就可以很快到不同的地方去玩。

Use conjunctions such as "首先", "而且" and "另外" to discuss the drawbacks of having your own car.

　　但是，自己开车也有坏处。首先，自己开车会破坏环境。因为汽车会排出废气，这些废气会提高地球的温度。现在我们的地球已经有全球暖化的问题，我不想让这个问题变得更严重。而且，汽车排出的废气会造成空气污染。另外，如果每个人都有自己的车，那么上班和上学的时候，路上的车多了，就会很容易堵车。

Reiterate your concern and ask for suggestions.

　　我很想要一辆自己的车，但我觉得每个人都有责任保护环境。你觉得我应该买车吗？我很需要你的意见，期待你的回复。

　　祝好！

朋友
玛丽

(357 words)

3 申请信

文体介绍

申请信是为了申请某一工作职位而写的信件。

申请信的内容：
- 申请的事项
- 申请的理由或目的
- 个人的能力说明
- 个人的诚意和希望
- 个人的联系方式

申请信的语气需谦虚客气。

3 Application letter

Text type introduction

The purpose of an application letter is to apply for a certain job position.

Content of an application letter:
- The position applying for
- The reason or purpose of application
- Abilities of the applicant
- Expression of the applicant's sincerity and hope
- Contact information

The tone of the application letter should be modest and polite.

格式 Format

（标题）

尊敬的负责人 / 领导 / 校长 / 老师：

（介绍自己来历，直接说明自己想申请的内容）

1、您好！我是来自XX学校的XXX，/ 我叫XXX，是来自XX学校的学生，我这次写信给您是想申请……一职。我愿意用自己的微薄之力，为……做出贡献。

2、我是来自XX的XX，此次来函是为了申请……一职 / 加入……。我愿意用自己的微薄之力，为……做出贡献。

（说明自己对于申请的工作的认识）

……工作是非常重要的。正是有了……的努力，……才……。

（说明自己适合的理由）

我相信自己能够胜任这份工作。/ 我觉得我有能力做好这份工作。

首先，……。

其次，……。

再者，……。

（感谢对方，再次表明希望得到这个工作的决心，并留下联系方式）

非常感谢您花宝贵的时间阅读我的申请信。希望您能给我面试的机会，我真心地期待与您的合作。若承蒙赏识，可以随时通过电邮或电话联络我。我的电邮是xxx@email.com，电话是2577 7844。

祝
工作愉快!

申请人

姓名

X年X月X日

范文 Sample essay

你从海报上知道学校正在为社区残障儿童招募义工，你很想参加这个活动。请写一封申请信给负责人，告诉他你为什么适合做这个工作。（HL）

Title ——— 申请社区残障儿童义工

尊敬的负责人：

Use "honorable" to show your respect.

　　您好！我是来自十二年级的陈小红。从学校海报得知你们正在为社区残障儿童招募义工，本人很感兴趣，也相信自己有能力担任这份工作。

Greet the recipient and state the position you are applying for.

　　本人乐于助人，相信"助人为快乐之本"，故自小就参加过不少的慈善活动，也有丰富的义工经验。在帮助别人的过程中，我获得了很多快乐，同时，也对人生有了更深入的了解。我曾到医院探访病童，与他们玩游戏，教他们英语。从这些经历中，我学会了怎样和小孩子相处，也知道自己很喜欢小孩子。

Briefly talk about your personality, work experience and qualifications.

　　其次，本人很有耐心。残疾儿童跟正常儿童不同，他们活动能力有限，但同样活泼好动。跟他们相处需要更多的爱心和耐心，还要时时关注他们会不会受伤。此外，残疾儿童在成长的过程中不免受到歧视。本人会格外小心谨慎，注意自己的言行，也会提醒其他队员，避免说一些可能伤害残疾儿童自尊心的话。

　　除此之外，本人持有有效的急救证，具备一定的医疗知识与实践经验。如果不幸发生意外，本人可以及时为伤者清洗及包扎伤口，甚至为伤者进行急救。

　　感谢您抽空阅读本人的申请信，希望您能考虑本人的申请，如需面试，请致电2577 7844。

Express your sincerity and provide your contact information.

　　祝

工作愉快！

Blessing: same as that of a letter.

申请人

陈小红

二零一四年三月四日

(426 words)

4 日记

文体介绍

把每天学习、工作、生活中的见闻有选择地、真实地记录下来的文字就是日记。

格式	首行一般写上日期、星期、天气情况。
内容	一般是记叙发生在自己身边的事情，用第一人称。
语气	语气不需要很正式，只要真实地记录下发生的事情和自己的心情便可。

4 Diary

Text type introduction

A diary is a type of personal writing written by an individual to record what has happened during the day. It may include the person's experience and feelings about study, work or life.

Format	The first line usually includes the date, day of the week and weather condition.
Content	Generally, the content is a record of what has happened around oneself, written in the first person perspective.
Tone	It does not need to be formal. Just record any daily events and your feelings in a casual or natural tone.

格式 **Format**

| X年X月X日 | 星期X | 天气：阴／雨／晴 |

今天我很开心，因为……

正文

从今天起，我打算／决定……

范文 Sample essay

你和你的好朋友大卫会去不同的大学读书，请写一篇日记，谈谈你的心情，并说说你将如何维持这份友谊。（SL）

二零一三年七月十日　　星期三　　雨 ──────

State the date, day of the week and weather condition.

Explain why you are upset.

我今天很不开心，因为我跟我最好的朋友——大卫将会去不同的大学读书。大卫会到美国读书，而我留在香港。

Recall memories of the relationship between you and your friend and express your feelings.

这几年我们天天见面，如果我们去了不同的大学就不能每天见面了，我一定会很想念他。以前我们每天一起上课，放学后一起去打篮球或复习功课。大卫的数学很好，每次考试都拿一百分，但是我的数学很差。所以，当我有不懂的数学题时，我一定会问大卫，他总能解释得很清楚。我真希望以后还能和他一起温习功课。

Describe how you will maintain the friendship.

我们上了大学后会认识不同的朋友，大卫还会记得我吗？为了继续保持这份友谊，我会在这个暑假约大卫出去玩。我要跟他看电影，去露营，去旅行。他去美国读书后，我每个星期都会给他发一封邮件，问问他那边的情况。我也可以每天上他的社交网站，看看他和朋友有什么趣事，我还可以给他打电话或者视频。现在的科技这么发达，只要我们都想念对方，就一定能保持这份友谊。我相信大卫跟我一样，无论去哪里，他都会想念我。

(357 words)

5 议论文

文体介绍

议论文又叫说理文，它是对某一事件进行评论或发表自己的观点、态度的文体。议论文三要素为：论点、论据、论证。

- 确定中心论点和分论点
- 结构要清晰（常见结构：开首——正文——结尾）
- 论据要充分
- 论证要有逻辑

5 Argumentative writing

Text type introduction

Argumentative writing shows a person's opinion or attitude toward an event. The three essential elements are argument, grounds of argument and demonstration.

- The argument and supporting arguments should be clearly defined.
- The structure should be clear. (The typical structure is: introduction, elaboration and conclusion.)
- The grounds of argument should be well supported.
- The demonstration should be logical.

格式 Format

（标题：论点）

（提出论点）

1、我认为……，理由如下：

2、……应不应该……，一直以来都备受争议。我认为……，理由如下：

（列点支持自己的观点）

首先，……

其次，……

再者，……

（总结强调自己的观点）

综上所述，我认为……。

范文 Sample essay

你学校的校报发出征文启事"学校是否应该规定和限制学生的穿着？"。请你写一篇文章投稿。（HL）

学校应该规定和限制学生的穿着

学校是否应该规定和限制学生的穿着？这个问题一直以来都备受争议，我认为：学校应该对学生的穿着有规定和限制。理由如下：

首先，限制学生的穿着能避免学生们穿一些不适当、不得体的衣服，提高他们的纪律性。规定与限制学生的衣着打扮就能避免不合礼仪的服装，以保证学生会穿着合适的服装。

其次，穿校服能够展现出学生的身份。就如社会里不同职业的人要穿不同的制服一样——消防员穿消防制服、警察要穿警察制服。同样地，学生也自然要穿校服。这样能让大家更容易、更清楚地分辨出学生的身份。

再者，穿校服可以消除学生互相攀比的不良风气。现在的学生强调个性、追求与众不同，有的学生还会利用服装去展现自我。比如说，有些学生爱穿名牌衣服——这不但会使其他同学感到自卑、沮丧，还可能会引起同学之间互相攀比。

此外，穿校服可以减少家庭的支出。对于一些低收入的家庭来说，如果儿女一星期五天上学都要穿不同的衣服，必定要花很多的钱，给家庭造成比较大的经济压力。

综上所述，无论从增加学生纪律性方面，还是减少花费等方面，规定学生的衣着打扮都有好处。所以，我认为：学校应该对学生的穿着打扮有规定和限制。

Use"首先"to state the first reason: to foster a sense of discipline among students.

Use"其次" to state the second reason: to show the identity of students.

Use"再者"to state the third reason: to prevent students from comparing themselves with others.

Use"此外"to state the fourth reason: to save money.

Conclude by restating the benefits of wearing school uniforms.

(440 words)

6 博客

文体介绍

博客，又译为网络日志或部落格等，是一种通常由个人管理、不定期张贴新的文章的网站。

格式	首行一般提供博客的名字或网站，然后写上标题和日期。
正文	鼓励读者留下意见：比如："大家对这个问题怎样看呢？欢迎给我留言"。
结尾	应列出：XX次阅读，XX个评论，XX收藏，XX转载，XX个回复等，最好包括其中的1—2个。

6 Blog

Text type introduction

A blog is also known as an online journal or online diary. A blog post is usually written by an individual and posted periodically.

Format	The blog name or website is usually given in the first line, followed by the title and date.
Body	Invite readers to leave comments, for example: "What does everyone think about this? Feel free to leave a message."
Ending	1-2 of the following should be included: Read xx times, xx comments, saved xx times, shared xx times, xx replies, etc.

格式 Format

范文 Sample essay

李大文博客

忽如一夜韩风来，韩剧韩娱中国开

Time of blog post ——— 2014年3月28日下午9时

不知从何时开始，韩流涌入了中国的大街小巷。我周围因"韩流""感冒"的大有人在，不论是00后少男少女，90后潮男潮女，还是80后俊男靓女，70后成男熟女，都是韩国电视剧和娱乐节目忠实粉丝，甚至中老年人中，也不乏痴迷疯狂者。

Describe a phenomenon.

博主有一朋友是韩剧迷。微博上的状态永远都是"（某某）偶吧，我爱你！"让人惊异的是她爱的"偶吧"如此多。忘了说明，"偶吧"是韩语对于男朋友或哥哥的称呼。对2000年以后的剧集，她几乎无所不知，无一不晓。她说，韩剧之所以吸引人，是因为打扮靓丽的明星实在养眼，满足了年轻人追求时尚的心理；也因为家境优越的男主角搭配出生平凡的女主角，童话般的爱情在现实生活中不易遇到。

Elaborate with examples.

如果说韩剧是中国人的老朋友，那么韩国娱乐节目则是新宠。某国内电视台买下韩国一档父子旅游的节目版权并翻拍成中文版，每到周五晚上10点，千家万户老老小小都守在电视机前，更有很多观众因此去搜寻韩国原版的来看。

Your personal views.

其实，韩剧韩娱表面看来只是形式，而韩国的生活方式，思维方式，文化传统都已经对国人产生潜移默化的影响。哎！热情地追逐别人的文化，自己老祖宗传下来的又在哪里？

关于韩流，大家是什么态度？欢迎给我留言！

Invite readers to respond.

52次阅读，15个评论

(467 words)

7 演讲稿

文体介绍

演讲稿又名演讲辞、讲话稿。一般用于在公众场合发表自己的意见，以达到某种目的。演讲可分为专题演讲和集会致辞。集会致辞包括介绍辞、欢迎辞、惜别辞、致谢辞等。结构如下：

称谓	一般场合：各位来宾
	学校：尊敬的校长，尊敬的老师，亲爱的同学们
	称谓的顺序：按照身份的尊卑（尊敬的校长、各位老师、各位同学）
开端	提出演讲的主题
	大家好！很荣幸今天在这里和大家分享……我演讲的题目是……
正文	有条理地提出自己的观点
	衔接词：首先……其次……最后…… 一方面……，另一方面……
结尾	概括自己的观点并致谢
	总而言之……
	综上所述……

7 Speech notes

Text type introduction

Speech notes are used in a public occasion to express a person's opinion for a particular purpose. Speech notes may be written on a specific topic or for an assembly. There are several types of speeches, such as an introductory speech, a welcoming speech, a farewell speech, a speech of acknowledgement, etc. Speech notes have the following structure:

Address	General occasion: Ladies and Gentlemen
	School event: Honorable headmaster, honorable teachers and dear students
	The order of address is based on the rank and position of the guests (honorable headmaster, teachers and students)
Opening	Introduce the topic of the speech
	Hello, everyone! It is my honor to be here today to share with you… The topic of my speech today is…
Body	Put forward one's opinions systematically
	Conjunctions: Firstly… Secondly… Finally… On the one hand… On the other hand…
Ending	Sum up one's ideas and express gratitude
	All in all…
	To sum up…

格式 Format

各位老师、各位同学：

　　大家好！今天很荣幸 ／ 高兴站在这里和大家一起分享我……的经历 ／ 对……的感受。我演讲的题目是《XXXXXXXXX》。 ／ 今天很荣幸（作为学生代表 ／ 代表……）在这里发言。

正文

　　首先，……

　　其次，……

　　再者，……

我真诚地希望……。 ／ 我相信，只要……，……就能……。

谢谢大家！

范文 Sample essay

你要参加一次中文演讲比赛，比赛的主题是"学习中文的重要性"。请你写这篇演讲稿。（SL）

各位老师、各位同学： ——— Teachers and fellow students,

大家好！我是中国国际学校的李大明。今天很荣幸能在这里和大家分享我对学习中文的看法。我演讲的题目是"这个时代学习中文的重要性"。 — Introduce yourself and state the topic of your speech.

我觉得这个时代学习中文是非常重要的。首先，现在中国的经济发展得很快，已经是世界上一个很重要的市场，很多不同国家的人都想到中国工作或与中国人合作。如果我们想得到好的工作，就应该学习中文，这样才能和中国人沟通。所以，学好中文可以使我们更容易找到工作。 — Give your first reason: it is easier to find a job if you speak Chinese.

其次，学会中文，我们可以看中文书，也可以利用中文在网上找中文资料。这样我们就能比不会中文的人获得更多的知识。 — Give your second reason: reading Chinese allows you to gain more knowledge.

此外，学中文可以使我们交到很多朋友。中国是世界上人口最多的国家，中文是他们最常用的语言。如果我们会说中文，就有可能认识很多中国朋友。中国有些小城市的风景很美丽，但是，那里的人英文都不太好。如果我们会说中文，就可以和那里的人交流，还能了解那里的文化。 — Give your third reason: you can expand your social circle and make more friends if you speak Chinese.

总而言之，我认为学习中文是非常重要的！
谢谢大家！ — Restate your topic.

(358 words)

8 建议书

文体介绍

建议书又称意见书，是个人或团体向其他人或组织就某一问题提出针对性的具体建议时所使用的文体，结构如下：

标题的写法有三种	XXXX建议书，例如：美化社区建议书
	建议XXXX，例如：建议美化社区
	关于XXXX的建议，例如：关于美化社区的建议
提出具体建议	实施方法
	预期效果
结束语	恳请认真考虑以上建议。

8 Proposal

Text type introduction

A proposal is a piece of writing in which a person or a group raises concrete suggestions about a specific issue to another person or organization. It has the following structure:

Three ways to write the title	"XXXX" Proposal, for example: "Community Beautification Proposal"
	Proposal on "XXXX", for example: "Proposal on Beautifying the Community"
	Regarding the proposal on "XXXX", for example: "Suggestions for Beautifying the Community"
Raise concrete suggestions	Implementation methods
	Expected outcomes
Ending	It would be highly appreciated if you would consider the above suggestions.

格式 Format

关于……的建议书

尊敬的XXX：

（提出建议的原因、理由）

（建议的具体内容）

第一、我认为……

第二、我们应该……

第三、我们还可以……

（提出自己希望被采纳的想法或意见）

……恳请考虑以上建议。

祝

工作愉快！

提建议的团体名称／个人的姓名

X年X月X日

范文 Sample essay

越来越多的人购买私家车，给环境造成了很严重的污染。请给《环境与世界》杂志写一篇建议书，说说车辆对环境造成的负面影响，并建议使用其他交通工具。（HL）

环保建议书

Introduce the topic of your proposal: how to reduce pollution.

尊敬的总编：

　　您好！城市里不少家庭都拥有两辆或以上的私家车，给环境造成极大的污染。我将在下面提一些我对环保的建议。

　　首先，使用私家车会严重降低城市的空气质量。众所周知，城市地少人多，高楼林立，汽车排出的二氧化碳把大厦围住，增加了空气中二氧化碳的浓度。市民如果吸入过多的二氧化碳，身体健康会受到严重的影响，甚至会引起中毒。

Use conjunctions such as "其次" and "此外" to put forward your opinions systematically.

　　其次，私家车也会排出大量热气，加剧温室效应，最后可能导致酸雨的形成。污染越严重，酸雨的酸度就越高，破坏力就越大。酸雨不但会破坏建筑物，还会伤害人类的身体。

　　此外，过多的私家车会造成交通阻塞。交通阻塞时，街上的车辆长时间排放二氧化碳和热气，使城市的空气质量进一步恶化。

Use point form to propose implementation methods.

　　因此，我们应该从现在开始改善城市的空气质量。对此我有以下建议：

1. 政府应该增加私家车的购买税，这样市民会少买私家车。
2. 政府应鼓励市民多乘坐公共交通工具，减少使用私家车，这样有助解决交通阻塞的问题。
3. 政府应该建造覆盖全市的单车径。

真诚地希望政府与市民能考虑本人的建议。

　　祝

工作愉快！

读者
黄大年
二零一四年三月十日

(407 words)

9 辩论稿

文体介绍

辩论稿就是在辩论中陈述自己一方观点，并做出简单解释的应用文体。

辩论稿要求：
- 结构清晰
- 立场鲜明
- 论据充足
- 运用口语化的语言

结构如下：

称谓	尊敬的主席、评委、对方辩友
正文	可用"首先、其次、最后"作为每一段的开头
结尾	综上所述……

9 Debate script

Text type introduction

A debate script is used by a debater to declare his or her standpoint with a brief explanation.

Requirements of a debate script:
- A clear structure
- A clear-cut stance
- Well-supported arguments
- Use of oral expressions

Below is the structure:

Address	Dear chairperson, judges and opposing speakers
Body	Use "firstly", "secondly"and "finally" to start each paragraph
Ending	To sum up...

格式 Format

尊敬的主席、评委、对方辩友：

大家好！我是正方的…… 我方的观点是：……（主题和观点）……。

首先，……

其次，……

最后，……

综上所述，……。所以，我方坚持认为：……（重复主题和观点）……。

谢谢大家！

范文 Sample essay

你要参加一个辩论比赛，题目为"学生应不应该追求最新的科技产品"。请选择你的立场，写一篇辩论稿。（SL）

尊敬的主席、评委、对方辩友： —— Formal greeting.

大家好！我是正方的李小珮。我方的观点是"学生不应该追求最新的科技产品"。 —— Firmly state your standpoint.

首先，拥有最新的高科技产品并不代表有知识。现在年轻人都有很多高科技产品，例如：手机、电脑、相机等。不少年轻人都有追求最新科技产品的习惯。每次有新款的手机发售时，他们一定会购买，因为觉得这是时尚。但是，有新科技产品，却不用来做有意义的事情，反而用来玩游戏、聊天。这样，不仅不能获得知识，反而会浪费时间和金钱。 —— Start with a topic sentence, followed by concrete examples and detailed elaboration.

其次，每天都有新科技产品出现，我们不可能追得上。现在科技发展快速，新的电子产品每几个月就会出现，我们不应该每几个月就换手机，这样太浪费金钱了。

最后，过分追求新科技产品会造成环境污染。如果我们每几个月就换一部手机或电脑，会制造很多电子垃圾。如果我们把旧的电子产品扔掉，这些产品含有的有毒物质，不但会污染环境，更会影响人类的健康。

综上所述，不追求最新科技产品既能避免浪费金钱，又不会污染环境。所以，我方坚持认为"学生不应该追求最新的科技产品"。 —— Conclude your argument: students should not blindly pursue the latest technology.

谢谢大家！

(378 words)

10 倡议书

文体介绍

倡议书是由个人或团体发出，用于宣传或推广某项活动的一种书信类型。书写格式与一般书信大致相同。结构如下：

标题	倡议书
	关于XXXX的倡议书
称呼	所倡议的对象，如"亲爱的同学们"、"各位老师"等
正文	倡议的目的和意义
	倡议的内容和具体事项
结尾	提出倡议者的希望或建议
署名	可为团体或个人
日期	年 / 月 / 日

倡议的内容需要很详细，包括活动的时间、地点、如何开展、具体要求等等。倡议的内容需要清晰明确、一目了然，因此可采用点列式将不同方面的内容分条目写出。

10 Proposal

Text type introduction

A proposal is a formal letter written by a person or a group to promote a certain event or activity. The format is similar to that of a general letter. It has the following structure:

Title	Proposal
	Proposal about XXXX
Address	The target group to which you are writing, for example, "Dear students", "Dear teachers", etc.
Body	The purpose and significance of the proposal
	The content and details of the proposal
Ending	State the writer's hope or further suggestions
Signature	The group or individual's name
Date	Year / month / date

Details should be given in the body, including the time, location, procedures and requirements of the activity. The content should be clear and easy to understand, with details listed in point form and separated into appropriate categories.

格式 Format

倡议书 / 关于XX的倡议书

各位同学 / 亲爱的同学们：

（倡议的目的、意义）

1、XXX同学不幸患上XXX病，由于他家境贫困，无法支付高昂的医疗费用。为此，学生会倡议全校同学，为XXX献出一份爱心，帮助XXX早日战胜疾病。

2、现在，环境污染越来越严重。为了保护我们的生存环境，我们倡议大家……。

（活动的具体事项）

1、捐款时间：……

　　捐款方式：……

2、我们应当做到以下几点：第一，……；第二，……；第三，……

学生会 / XX社团社长：XXX

X年X月X日

范文 Sample essay

你是学生会会长，请写一份倡议书鼓励同学们改善食堂浪费的情况。（HL）

减少食堂浪费倡议书

各位同学： ——— Dear students,

最近我们发现食堂的浪费情况非常严重，垃圾桶里都是同学们吃剩的饭菜和塑料餐盒、塑料餐具及纸杯等。这样做不仅会吸引很多苍蝇，还会使食堂的环境卫生变差。本会认为同学们应提高环保意识，减少资源浪费。在此，本会向同学们提出以下倡议：

Talk about the situation in the school canteen. Use "本会" to represent the student union.

第一，同学们如果食量小，应该向食堂服务员少要一点饭，以免造成浪费。如果那天的菜式不合胃口，也不应该因此而扔掉食物，应该先问一下其他同学愿不愿意和你交换，如果真的没有人愿意，也最好把饭菜吃完。要知道粮食来之不易，我们绝不能浪费食物。

Give detailed suggestions on how to avoid waste.

第二，同学们可以自备餐盒和食具，这样就不需使用一次性塑料饭盒和餐具。塑料是很难分解的化学物品，它在分解过程中会释放有毒气体，严重危害人类的健康，所以，我们应该减少使用一次性的塑胶制品。另外，虽然纸杯和纸巾都不是用塑料制造的，没有难分解的问题，但它们也是一次性的制品，非常不环保。同学们应该自备水瓶，或者使用食堂提供的玻璃杯，也可利用手帕代替纸巾。

Propose ways to reduce environmental pollution.

改善食堂环境，减少食堂浪费，就这么简单！希望同学从今天开始一起行动，保护环境！

Express your hope.

学生会

二零一四年三月七日

(424 words)

11 读后感 / 观后感

文体介绍

读后感、观后感就是读了一本书、一篇文章，或看了一部电影之后，把具体感受写成的文章。

结构如下：

标题	读《XXXX》有感 / 观《XXXX》有感
正文	有一个明确的中心论点（或中心思想）。
	这个中心论点必须是从所读的材料或观看的电影中概括出来的。
	写自己的感想。

11 Book review / film review

Text type introduction

A book or film review is written to express a person's feelings after reading a book or an article, or watching a film. It has the following structure:

Title	Review of the book "XXXX"/Review of the film "XXXX"
Body	Give a clear argument or theme.
	The argument or theme should be generated from the book or the film.
	Express one's own feelings.

格式 Format

读《XXX》有感 /《XXX》观后感

（引入话题）
读完这本书，我被……深深地感动了。（读后感）
电影《XXX》向我们讲述 / 展示了……的故事。（观后感）

（正文：介绍书 / 电影的内容或它给你印象最深的地方）

（结合自己的现实生活谈感想，总结全文。）

范文 Sample essay

你刚刚读完了一本书，书中的主人公让你很受启发。请写一篇读后感。（HL）

读《饥饿游戏》有感

Give a brief summary of the book. The last sentence of this paragraph connects with the following paragraphs.

最近，我读了美国作家苏珊柯林斯的书《饥饿游戏》。故事发生在一个虚构的时代，那个地方每年都要选出24个年轻人参加残酷的"饥饿游戏"。他们彼此相互残杀，最后只有一个人能活下来。主人公凯特尼斯在这个游戏中经历了重重磨难，体会了亲情、友情和爱情的酸甜苦辣，慢慢地成长起来。本书最吸引我的正是主人公的性格。

Elaborate on the heroine's characteristics in three paragraphs and start each paragraph with a short topic sentence.

她坚强。她自小失去了父亲，和妈妈、妹妹相依为命，成为家里的经济支柱。为了家人，她冒着被统治者抓到的危险到黑市把猎物卖掉。当她的妹妹波丽姆被选去参加饥饿游戏时，她义无反顾地选择了亲情，自告奋勇代替妹妹。

她善良。她和竞争对手露露结下了深厚的友谊。当露露被其他竞争者杀死后，凯特尼斯为她报仇，并用鲜花和歌声为她装点生命的最后一程。

她勇敢而反叛。她清楚地知道自己不想被游戏组织者控制，于是利用比赛的漏洞，为自己和搭档赢得了生存的机会，后来更勇敢地承担起反抗暴力统治的重任。

Express your feelings and talk about what you have learnt from this book.

读完这本书以后，我非常佩服这个女孩。她让我明白，机遇是留给有准备、有勇气的人的。同时，这个故事也让人深思战争的残酷。我希望世界和平，永远不会发生《饥饿游戏》中的战争、强权和厮杀。

(428 words)

范文 Sample essay

你最近看了一部电影，觉得很精彩，写一篇观后感。（HL）

电影《千与千寻》观后感

《千与千寻》是一部百看不厌的好电影。它的故事很有趣，画面色彩缤纷，音乐也很动听，而且故事中出现的人物和背景展现出日本的传统文化，非常引人入胜。

Briefly introduce different aspects of the film.

这是个充满想像力的故事：在迁往新家的途中，千寻和父母无意间闯入了神秘的小镇，里面有很多美食。千寻的父母因为贪食而被变成了猪，虽然千寻非常害怕，但她为了拯救父母，便勇敢地到了邪恶女巫汤婆婆的"汤屋"打工。

Talk about the opening of the film.

刚到"汤屋"的时候，千寻被改名叫作小千，白龙让她记住自己原来的名字，因为如果她忘记以前的名字，便无法回到原来的世界。在"汤屋"中，千寻还遇到了河神。当他来"汤屋"洗澡时，全身肮脏，挂满垃圾，散发出腐烂的味道。大家都把他当成了腐烂神，远远地躲开。只有千寻和大家不一样，她为河神清理污垢，最后终于帮他清洗得非常干净。这是一个引人深思的画面，江河本来是很干净的，为什么在这个电影里那么脏呢？其实，这反映了人们对江河的污染，对自然的破坏，强调了环保的重要性。

Talk about the main story and the most memorable scene in the film.

我喜欢这部电影，它让我体会到了人与人之间的真善美；它让我看到一个女孩由娇宠、懒惰、胆小到坚强勇敢的蜕变；它更让人们思考我们应该如何与自然相处。

Sum up your feelings and say why you like this film.

(424 words)

12 注意事项

文体介绍

注意事项是单位或个人为了提醒或强调应该注意的事情而写的一种应用文体。

语言简明扼要，言简意赅，不应过于复杂。结构如下：

标题	XXXX注意事项
称谓	"各位居民"或"各位同学"
内容	列点写出
署名	可为团体或个人
日期	年 / 月 / 日

12 Notice

Text type introduction

A notice is usually written by an organization or an individual to remind others of something or to emphasize the importance of specific matters.

The language should be brief and straightforward. Avoid using complicated expressions. It has the following structure:

Title	Notice on XXXX
Address	"Residents" or "Students"
Content	In point form
Signature	The group or individual's name
Date	Year / month / date

格式 Format

XXXX注意事项

各位同学：

（交代一下情况，或者写这个注意事项的原因）

正文分点列出要注意的事项

1、

2、

3、

4、

5、

希望大家注意以上事项……

XX部／XX学会／XX学生会

X年X月X日

范文 Sample essay

在你住的地区发生了一种流行病。你作为学生会主席，写一份流行病预防指南，告诉同学们这种流行病的注意事项。（SL）

预防流行感冒注意事项

各位同学：

最近我们住的地区有很多人感冒了。学生会提醒同学们要好好爱护自己的身体。为了避免生病，请注意以下事项：

1、要多吃新鲜的蔬菜和水果，特别要多喝热水。

2、不要吃太多不健康的食品，如：快餐、油炸的东西、辛辣的和太冷的食物。

3、要有充足的休息时间，早睡早起。

4、去洗手间之后或吃东西之前一定要洗手。

5、如果觉得头痛或者发烧，应该立即去看医生。

6、咳嗽时一定要用纸巾捂住嘴巴和鼻子，用过的纸巾要立即扔进垃圾桶里。

7、生病的同学最好留在家里好好休息，不要上学，这样就不会传染给其他同学。

8、感冒的同学一定要按时吃药，这样才可以早点康复。

9、不要去人多拥挤的地方，如果一定要去，最好带口罩，防止细菌进入我们的身体。

另外，学校医疗室已准备了一些预防感冒的药物，如果同学们有需要，可以到五楼的医疗室去拿。

希望大家注意以上事项。

学生会
二零一三年十月四日

Talk about the current situation and remind students to protect themselves against disease.

List the measures how students can protect themselves.

Add any other relevant information.

(338 words)

13 介绍性文章

文体介绍

介绍性文章是对某一事物、地点或人物做出客观、简单的介绍。

介绍的语言要求准确、客观、简炼。介绍的手法可以使用列数字、作比较、下定义、打比方、举例子、分类别等。

结构通常分为三部分：

开端	开头引出介绍物件
正文	详细解说介绍的物件
结尾	总结介绍的物件

13 Introductory article

Text type introduction

An introductory article gives an objective and brief introduction to an event, place or person.

The language should be precise, objective and brief. The article can be written by listing concrete numbers, comparing, defining, drawing analogies, giving examples, categorizing, etc.

The structure of an introductory article consists of three parts:

Opening	Briefly introduce the subject
Body	Describe the subject in detail
Ending	Sum up the characteristics of the subject

格式 Format

（标题）

（简单介绍该事物）

（分段或列点介绍该事物的主要内容）

（概括该事物特点，或呼吁大家来购买、欣赏、体验）

范文 Sample essay

你的朋友对中国菜式很感兴趣，给他介绍一道中国的名菜。（HL）

中国名菜——北京烤鸭

北京烤鸭是驰名中外的中国名菜之一，是国内外游客到北京旅行时不可错过的美食。

其实，北京烤鸭本来是南京的特产，它一开始只是一般南京市民会吃的菜式。但是，因为太好吃了，而且太出名了，所以在古代，连皇帝都对它赞不绝口，最后传到了北京。

以前的中国人很喜欢吃肉，北京烤鸭也是越肥越好。后来，人们为了健康，就改用比较瘦的鸭子了。北京烤鸭的做法比较复杂，现在让我来介绍一下吧！

首先，把鸭子的鸭掌和鸭舌都去掉，把内脏取出来，然后一直洗，直到洗得非常干净。之后，要把鸭子挂起来，用热水把鸭子的皮都弄熟，然后再把糖水涂在鸭皮上，这样鸭皮会变得又甜又脆。等鸭子完全风干后，就可以把水放进鸭子的肚子里，水可以令鸭子的内部冷一点，这样鸭肉会更嫩。接着，把鸭子放进烤炉，烤到鸭子变成金黄色。最后，把烤好的鸭子切成小薄片。此外，吃的时候，要把烤鸭、葱和甜酱用薄饼包起来一起吃，这样味道最鲜美！

相比起自己在家做北京烤鸭，更多人喜欢去餐厅吃。而在北京，最有名的北京烤鸭餐馆在王府井。如果想吃正宗的烤鸭，一定要去那儿尝尝！

(402 words)

Briefly introduce the dish (i.e. Beijing roast duck).

Talk about the history of the Beijing roast duck.

Explain the cooking method in detail.

Tell readers where they can try the dish.

14 操作步骤

文体介绍

操作步骤就是将完成某一任务的步骤顺序记录下来的文体。常见的例子有菜谱制作说明。结构如下：

标题	"XXXX的制作方法"或"自制XXXX"
正文	正文部分应列点写出。注意连接词的使用：首先，……
	接着，……
	然后，……
	之后，……
	最后，……
结尾	如有需要，应写明制作工具或原材料。如有一些注意事项，也应在最后列明。

14 Operation procedures

Text type introduction

An operation procedure records the steps to perform a specific task. Some common examples are recipes and instructions for production. It has the following structure:

Title	"XXXX Production Methods" or "Homemade XXXX"
Body	The body should be in point form. Use connectives whenever possible: Firsty, XXXX
	Then, XXXX
	Moreover, XXXX
	After that, XXXX
	Finally, XXXX
Ending	List the tools or raw materials needed if applicable. If necessary, list the matters that need special attention at the end.

格式 Format

XXX的制作方法

制作材料：

制作工具：

制作步骤：

    1、……

    2、……

    3、……

    4、……

    5、……

注意事项：

    1、……

    2、……

    3、……

    4、……

范文 Sample essay

你朋友想学做中国菜，请写一道菜的菜谱，介绍一道中国菜的制作方法。（HL）

可 乐 鸡 翅 的 制 作 方 法

可乐鸡翅是一道常见的菜。这道菜有许多种做法，但主要材料是鸡翅和可乐。可乐鸡翅味道鲜美，既保留了鸡翅本身的嫩滑，又吸收了可乐的香味，深受人们的喜爱。其具体的制作过程如下：

制作材料：8只鸡翅、1小勺味精、1杯可乐、1/4杯酱油、1大匙糖、2根葱。

制作步骤：

1、把可乐、酱油、味精、糖混合成腌料。

2、用小刀把鸡翅切开一道口子，以方便腌料渗入。

3、把鸡翅浸入腌料中，腌30分钟左右。

4、鸡翅腌好以后，先倒少许油到锅里开至大火，等油热以后转小火，再把腌好的鸡翅放进锅里煎至6成熟。

5、把鸡翅捞起，和腌料一起放入另一锅中，并用大火将其煮滚。

6、转小火煮半个小时，直到鸡翅变成棕色。

7、熄火并放上葱即可。

注意事项：

1、腌鸡翅过程中，要确保鸡翅全部浸在腌料里。可不时用筷子翻动鸡翅，让腌料全面渗入，但不必不停搅拌。

2、将腌好的鸡翅放下锅时注意一个一个轻轻放，避免热油弹出，烫伤皮肤。

3、煎鸡翅时必须使用小火，防止火势过大将鸡翅煎糊。

4、鸡翅和腌料一起煮时，要多加留意，防止汁水煮干，鸡翅粘锅。

> Introduce the "cola chicken wing" briefly.

> List the ingredients and utensils needed. Students should pay attention to the units of measurement.

> Use point form to list out the steps.

> Give any additional tips or reminders.

(384 words)

15 产品介绍

文体介绍

产品介绍是产品生产者向消费者介绍其产品性能、特点、作用等基本信息的一种文体。常见的例子有广告、宣传单和小册子。结构如下：

标题	用产品的名字做标题
正文	可以分段介绍产品不同方面的特点，也可列点写出产品特点、性能。

若作为广告，可适当加入修辞手法。如比喻、夸张、拟人等。

15 Product information

Text type introduction

Product information is written by producers for consumers to provide product information, including its nature, characteristics and functions. Some common examples are advertisements, leaflets and booklets. It has the following structure:

Title	The product name
Body	Write in paragraphs to describe different aspects of the product or list in point form its characteristics and functions.

In an advertisement, rhetorical devices, such as metaphor, exaggeration, personification, etc. can be used.

格式 Format

（标题）

（简单介绍该产品）

（介绍产品的各项特点：分段或列点）

（概括产品的特点）

范文 Sample essay

你是一家科技公司的职员，最近你们公司为年轻人设计了一款新产品。经理让你为这个新产品写一个小册子，宣传产品的功能特点。（HL）

创新笔记本型电脑

Describe the product briefly.

创新笔记本型电脑是本公司今年专门为年轻人设计的新产品。它外形美观、功能齐全，受到时下年轻人的追捧。它令人喜爱的原因多不胜数：

List the reasons why the product is suitable for young people.

第一，它的外形非常漂亮。有白色、黑色、橙色、紫色、红色、蓝色、黄色等七种颜色，每种颜色还配有不同的时尚图案，十分符合年轻人爱美的个性。

第二，它非常轻便。它不但外形小巧，放在哪里都不占地方；而且只有1千克重，方便携带，每天把它放在手提包里外出也很方便。

第三，它的电池比一般的电池耐用。即使您在没有电源的地方使用，它也可以一直工作大约十个小时，帮助您完成繁重的工作。

第四，它还能保护您的眼睛。由于现代年轻人每天会有大量时间使用电脑，因而他们的眼睛易疲倦，也容易受到伤害。为此，本公司为创新笔记本型电脑设计了护眼的功能，让大家长时间使用电脑之后也不会感到眼睛疲劳。

第五，它有着电影院般的画面和声音效果。就算在家里用创新笔记本型电脑看电影，也有在电影院看电影的效果，这样一来，您以后都可以在家里看电影了。

Start the paragraph with a question to attract readers.

怎么样？拥有这么多优点的创新笔记本型电脑是不是很吸引您呢？心动不如行动，赶快把它带回家吧！

"心动不如行动" can be used as a slogan in advertisements.

(417 words)

16 活动介绍

文体介绍

活动介绍是主办单位向公众或团体告知举行各类活动的一种应用文体。常见的例子有海报、传单、广告。结构如下：

标题	单独由文类构成。即在第一行中间写上"海报"字样。
	直接以活动的内容作为题目。如"影讯"、"球讯"等。
	使用一些描述性的文字。如"XXXX再显风采、XXXX旧事重提"。
正文	活动的目的和意义。
	活动的主要内容、时间、地点等。
	参加的具体方法及一些注意事项等。
落款	主办单位的名称及发文日期。

16 Activity introduction

Text type introduction

An activity introduction is usually written by an organizer to announce to the public or a group of people a certain activity. Some common examples are posters, flyers and advertisements. It has the following structure:

Title	The name of the text form, such as "Poster", written on the first line.
	The name of the activity, such as "Movie News" or "Sports News", etc.
	Use descriptive language.
Body	The purpose and meaning of the activity.
	Events, time, location, etc. of the activity.
	Enrollment details and any other special remarks.
Signature and date	Name of the organizer and the date of releasing the text.

格式 Format

（标题）

活动的目的、意义或举办原因。

活动主要内容：

    1、

    2、

    3、

报名方法：电话：2577 7844；电邮：xxx@email.com

活动时间：X月X日 X时X分

活动地点：

费用：每人XX元

注意事项：（可以不写这一部分）

    1、

    2、

    如有任何查询，请致电2577 7844。

（主办单位名称）

X年X月X日

范文 Sample essay

你是学生会主席，打算组织同学到地震灾区做义工。请写一篇海报宣传这个活动。（SL）

四川灾民正等你伸出援手 ——— Title: summarize the poster in one sentence.

Describe the situation after the earthquake in Sichuan Province.

———— 五月十二日，四川发生八级大地震。到目前为止，已有两百多人死亡，六百多人受伤。当地学校全面停课，居民的房子都倒了，数以千计居民无家可归。大部分的居民都睡在街上，他们缺水缺食物，生活条件非常差。

State the purpose of the activity: to help victims in Sichuan.

———— 同学们，当你们正在看这张海报的时候，你们身在美丽的校园，正与亲切的老师或亲爱的朋友享受着美味的小吃和饮料。这个时候，你有没有想过四川的学生，他们失去家园、父母、老师及朋友，有些更失去手脚，甚至丧失了生命。他们需要我们的帮助！

List the main content of the activity.

———— 为了使同学们更关心社会，帮助他人，我校学生会将于6月15日至6月22日组织同学到四川做义工。活动内容如下：

1、把全校捐的钱和物品运到四川发给灾民。

2、帮助灾民搭建房子。

3、给那里的学生上课。

4、帮助医生照顾病人。

Provide any other relevant information, e.g. how to join, registration period, fee, contact number, etc.

报名方法：致电2577 7844或电邮至xxx@email.com

报名时间：即日起至6月13日

费用：每人￥1,500

如有任何查询，请致电2577 7844。

学生会

二零零八年六月一日

(341 words)

17 新闻报导

文体介绍

新闻报导亦叫做新闻稿，是对最近发生的事情的报导。

写作要素	时间、地点、人物、发生的事情
语言特点	准确、客观
写作方法	符合事实
	列出具体资料或引用原话
语气	最好选用第三人称，比如："据记者采访得知……"。

17 News report

Text type introduction

A news report is a report of the latest events that happen around us.

Writing elements	Time, location, character, event
Language	Precise and objective
Writing techniques	Reflect the truth
	List concrete data or use direct quotes
Tone	Use narrative tone and third person, such as "Through an interview, the reporter realized that…".

格式 Format

（标题）

（概括新闻事件的时间、地点、人物、发生什么）

本报讯，……

（对新闻事件做具体的报导，可引用原话或具体数字）

（总结新闻事件）

范文 Sample essay

你的学校刚举行了春季运动会。作为校报记者,请你写一篇新闻报导。(SL)

学校春季校运会创下新纪录

本报讯,4月29日至4月30日,本校春季运动会在体育馆举行。这次运动会不仅在项目、参赛人数上比以往有所增加,很多项目还创下了本校运动会的新纪录。

本次运动会共有比赛项目15个,比去年增加了2个。新增项目为室内游泳和网球,吸引了许多同学去现场观看。据记者了解,这次运动会共有师生200人参加,比去年增加了50人。另外,还有10名教师和20名同学担当本次运动会的裁判。除了比赛项目和参与人数打破了本校的纪录以外,这次运动会的比赛结果也打破了多项纪录。特别是男子100米跑步,王明同学以11秒零5获得了冠军。这个成绩打破了本校的男子100米短跑纪录。

昨日下午,随着女子接力赛的结束,东华国际学校春季运动会也圆满结束。张校长接受本报记者采访时表示,全体师生在这次运动会中充分发扬了团队精神和体育精神,大家的表现非常不错。

Use "本报讯" as a third person. Introduce some basic elements: time, location, people involved and event, etc.

Talk about the sports games in detail.

Conclude the event with the interview with the principal.

(328 words)

18 人物访谈

文体介绍

人物访谈又叫人物专访或采访稿。人物访谈是为某个特定话题，去访问一些专家、著名人物或者事件相关者，请他们对提出的问题进行解答，并将访问过程记录下来的文章。

采访稿可以分为两种：访谈式和叙述式

- 访谈式采访稿的组成部分：题目、开头（比如访问的引言或者时间、地点、人物等资讯）、对话内容（一问一答）、结尾（简要总结采访内容）。访谈的写作要力求全面记录与访谈人物的谈话内容。
- 叙述式采访稿的组成部分：题目、开头（引言）、主干、结尾（简要总结采访内容或者表达记者个人感受）。

采访稿的标题：

主标题	应根据采访主题拟定
副标题	可用"访 + 称呼 + 人名"的格式 如："访著名歌手王海"

18 Interview

Text type introduction

An interview is also called an exclusive interview or interview draft. It is usually written in the form of a conversation that reflects the interviewees' views or suggestions on a specific topic. Interviewees are usually experts, famous people or people who are related to the event reported.

An interview report can be a transcription of the dialog or presented in a narration format.

- A transcription report includes: title, opening (e.g. brief introduction, time, venue, interviewee, etc.), the dialog in question and answer format, ending (summarize the interview). The interview should be a full record of the conversation with the interviewee.
- A narrative report includes: title, opening (introduction), main body, conclusion (summarize the interview or express one's feelings of the interview).

Title of the interview:

Title	Based on the theme of the interview
Subtitle	"Interview with + address + name of the interviewee" For example: "Interview with Singer Wang Hai"

格式 Format

（主标题）
——访XXX／XXX专访

（简单概括人物、事件）

（非问答式：详细介绍事情的内容／问答式：记者与被采访者的对话）

（结尾）

范文 Sample essay

学校来了一位新老师，用问答形式写下你跟他／她的访谈内容。（HL）

Title: "A person who spreads Chinese culture"

中国文化的传播者
—访问李老师

Use a question to start.

　　最近有一位新的中文老师加入了我们学校，但是大家对她的认识又有多少呢？我们今天很高兴能采访到这位神秘的李老师，让同学们更了解她。

The interview in question and answer format.

记：李老师好！您为什么会选择教中文？

李：我从小就喜欢中文和中国文化，而且也喜欢跟你们这些年轻人沟通。跟你们在一起，我觉得自己也年轻了。此外，现在的年轻人不是很重视中国传统文化，反而受西方文化影响，我担心有一天中国传统会被淘汰。所以，我希望能把传统文化的精华传授给学生，让他们明白中国传统并不代表落后，也有可取的地方。

记：您觉得我们学校的学生中文水准怎么样？

李：学生们的中文水准一般，他们的英文比中文好。但是，只要上课专心，并且按时完成作业，相信他们的中文水准会有很大的进步。

记：您上课时遇到的最大困难是什么？

李：是中文程度的差异吧！学生拥有不同的背景，有些来自中国家庭，中文水准固然比较好；有些来自国外，除了中文课，不会接触中文，程度当然没那么好。我课上会先用中文说一遍，然后再用英文解释，务求让所有同学都明白。

　　李老师是一位和蔼可亲、因材施教的老师，如果大家在中文上有什么问题，可以请教她。

(422 words)

范文 Sample essay

一位有名的歌星到你的学校做慈善表演。你作为校报的记者对他／她做了访问。请写一篇访谈稿。（HL）

成功、机会、勤奋
——访著名歌星林大海

　　昨天，华语乐坛知名歌星林大海来到我校，为慈善晚会做表演嘉宾。当晚，他精彩的演出把全场气氛推向高潮。晚会结束后，我们有幸对他进行了访问。

　　林大海三年前参加全国歌唱比赛获得冠军，因此得到人生中的第一张合同，成为耀星音乐公司力捧的歌手。他将自己的成功归功于勤奋与毅力。他说，他认识的许多人都很有音乐才华，但他们总认为自己是因为没有机会才没能成功，而忽略了"机会只留给有准备的人"。实际上，没人知道机会什么时候会来，但是准备却可以随时做。如果没做好准备，机会来了也没办法把握，那时就只能追悔莫及。林大海说自己不是一个等待机会的人，他觉得机会是自己争取来的，不是"等"来的。正如他成为明星前，曾参加超过五十场歌唱比赛，虽然大部分都没获奖，但他没有放弃，而是从失败中汲取教训，积累经验，再接再厉。正是靠着这股精神和毅力，林大海才一步一步、踏踏实实地迈向成功。

　　林大海很欣赏我校能够举办慈善活动，关心弱势社群。他每月都会把一半的薪金捐给慈善机构，希望给更多人带来希望，带来机会。

　　最后林大海鼓励同学们，要给自己完成梦想的机会，也别忘了给别人完成梦想的机会。

(436 words)

Title: "Success, Opportunity and Hard Work Interview with Singer Lin Dahai (林大海)"

Talk about the background of the interview.

Summarize the main content of the interview and talk about Lin Dahai's (林大海) road to success.

End the interview with Lin Dahai's (林大海) encouragement to students.

Option topic 1

Cultural diversity

选修主题 1

文化的多样性

1.1 人体美的概念
Concept of physical beauty

关键字 Key words

外表	身材	打扮	皮肤	高挑
性感	时尚	出众	漂亮	吸引

模拟试题 Mock questions

SL	1. 每个人心中美的概念都是不一样的，你觉得什么是美？
	2. 谈谈什么是"人体美"。
	3. 外在美和内在美哪个更重要？
	4. 有些人觉得瘦就是美，说说你的看法。
	5. 很多年轻人减肥，谈谈你的看法。
HL	1. 你对现在年轻人减肥有什么看法？
	2. 谈谈现今社会整容的风气。
	3. 谈谈你对现今年轻人追求时尚的看法。
	4. 谈谈你对外在美和内在美的看法。
	5. 谈谈穿着打扮与美的关系。

范文 Sample essay

你觉得什么是美？青少年应该追求怎样的美？（SL）

我眼中的美

　　每个人对美的看法都不一样。有些人认为"美"就是拥有漂亮的外表和高挑的身材，所以他们不仅经常打扮自己，还要减肥，让自己有苗条的身材。的确，如果看见一个样子出众、身材高挑，而且穿着时尚的女孩，每个人都会觉得她很美丽。但是，"美"就只是这样吗？

Talk about people's general perceptions of beauty.

　　什么是"美"呢？我认为美分为两种：外在美和内在美。其中外在美就是前面所说的外表、身材和打扮。外在美虽然能吸引别人的目光，但是不一定能让别人喜欢你。真正的美丽应该是内在美，内在美包括良好的个性、善良的心地和丰富的学识。有些人既个性开朗又乐于助人，我觉得他们是美丽的，因为他们又自信又善良。还有些人虽然样子普通，但是有丰富的学识，我觉得他们也是美丽的，因为他们有内在美。

Define outer beauty and inner beauty. Emphasize that true beauty is what lies inside.

　　在我看来，我们可以追求外在美，因为拥有外在美是一个人的优势。但是，我们不能只追求外在美，而应该更注重自己的内在美。我们应该努力让自己成为一个自信、善良和有内涵的人。

Conclude the passage by encouraging readers to focus more on inner beauty.

(347 words)

词汇 Vocabulary

每个人	everyone		其中	among them
看法	idea		虽然	although
都	all		吸引	attract
不一样	different		别人	other people
有些人	some people		目光	attention
认为	think		但是	but
拥有	possess		不一定	may not
出众	outstanding		真正	real
外表	appearance		应该	should
高挑	slender		包括	include
身材	figure		良好	good
所以	so		个性	personality
不仅	not only		自信	self-confidence
经常	often		善良	kind
打扮	dress up		丰富	a wealth of
自己	oneself		学识	knowledge
减肥	lose weight		开朗	cheerful
的确	indeed		乐于助人	helpful
如果	if		觉得	think
穿着	outfit		优势	advantage
时尚	fashionable		注重	pay attention to
美丽	beauty		内涵	inner quality
分为	divide into			

语法 Grammar

1. 不仅……还 / 而且 (not only... but also)

The construction indicates a progressive relationship. The usage is the same as "不但……而且……".

Examples:

有些年轻人不仅经常打扮，还刻意减肥。

Some young people not only dress up, but also lose weight deliberately.

他不仅学习好，而且体育也好。

He is not only good at studying, but is also good at sports.

2. 虽然……但是 (although)

The construction is used to show an adversative relationship. "虽然" may go either before or after the subject.

Examples:

外在美虽然能吸引别人的目光，但是不一定能让别人喜欢你。

Although physical beauty may attract people's attention, it may not necessarily make people like you.

他虽然学中文的时间不长，但是学得很好。

Although he hasn't been learning Chinese for a long time, he is making good progress.

3. 的确 (really; indeed)

This adverb can be placed at the beginning of a sentence or after the subject to emphasize a statement.

Examples:

的确，他的小提琴拉得很好。

Indeed, he plays violin well.

我的确喜欢在图书馆看书。

I really like reading in the library.

词组及常用句式
Expressions & sentence structures

词组 Expressions

对美的看法	perception of beauty
有些人认为	some people think that
漂亮的外表	a beautiful appearance
高挑的身材	a slender figure
打扮自己	to dress up oneself
苗条的身材	a thin figure
样子出众	an attractive appearance
穿着时尚	fashionably dressed
只是这样	just so
吸引别人的目光	attract the attention of others
真正的美丽	true beauty
良好的个性	a kind personality
善良的心地	kind-hearted
丰富的学识	a wealth of knowledge
个性开朗	a cheerful character
乐于助人	willing to help others
自信又善良	full of self-confidence and kind
在我看来	in my point of view
注重自己的内在美	focus on one's inner beauty

常用句式 Sentence structures

1. (sb.) 对 (sth./sb.) 的看法

Explanation	Example
somebody's view of something	每个人对美的看法不一样。 People have different views of beauty.

2. (sb.) 认为 / 觉得

Explanation	Example
somebody thinks/ considers that	我认为内在美才是真正的美。 I think that real beauty is what lies inside.

3. (sth.) 分为A (sth.) 和B (sth.)

Explanation	Example
something is divided into A and B	美分为外在美和内在美。 Beauty can be divided into physical beauty and inner beauty.

4. (sth.) 包括A (sth.) 和B (sth.)

Explanation	Example
something includes A and B	外表包括样子、身材和打扮。 Outer appearance includes a person's look, figure and way of dressing.

范文 Sample essay

你对现在年轻人减肥有什么看法？（HL）

谈谈现今的纤体减肥潮流

Briefly describe the weight loss trend.

现今纤体减肥的风气弥漫，纤体减肥的广告随处可见。翻开报纸，一边是减肥广告，另一边却是青年女子因节食晕倒的新闻。

Compare how beauty is perceived in the Tang dynasty and in modern times.

对瘦的崇尚已经成为一种潮流，而带动潮流的是人们日渐改变的审美观。中国唐朝的时候，人们认为胖的女子才是最漂亮的。但如今，人们以瘦为美，几乎人人都追求苗条的身材。

Define "beauty" and elaborate by giving examples. Emphasize that beauty is not just merely about outer appearance; it is more about one's inner qualities.

"瘦"真的是"美"的唯一标准吗？其实，"美"不能用身材来衡量。无论你是胖还是瘦，都能拥有美丽。因为美并不仅仅限于外表，还有很多其他的因素，比如性格、知识和修养。有的人虽然天生容貌靓丽，身材高挑，但是却性格高傲，只追求物质，不在乎自身的修养。这样的人算得上"美"吗？相反，有些人没有别人眼中姣好的身材，但却有丰富的学识和良好的修养。这样的人气质出众，拥有令人敬佩的内在美。所以，美不美不只是在于外在的容貌和身形，更多的是一个人的修养和气质。

Talk about the negative influence of the weight loss phenomenon and remind readers not to become victims of this unhealthy trend.

虽然爱美之心人皆有之，但现今的纤体减肥潮流却促使人们过度追求外在美。这不但会导致不合理的价值观，让年青一代盲目纤体减肥，而且极有可能对他们的身体健康造成伤害。因此，我们一定要认清什么是"美"，不要当"纤体"的奴隶。

(416 words)

词汇 Vocabulary

纤体减肥	keep fit and lose weight
风气	trend
弥漫	filled with
广告	advertisement
随处可见	omnipresent
翻开	flip open
报纸	newspaper
青年女子	young lady
节食	on a diet
晕倒	faint
新闻	news
崇尚	advocate
已经	already been
成为	become
潮流	fad
日渐	gradually
改变	change
似乎	seems to
唯一	the only
标准	standard
其实	in fact
衡量	judge
因素	factor
比如	for example

修养	accomplishment, cultivation
天生	inborn
容貌	appearance
高傲	proud
物质	materialistic
自身	oneself
相反	on the contrary
姣好	charming
气质	refinement
敬佩	admire
概念	concept
促使	impel
导致	lead to
不合理	unreasonable
价值观	value
年青一代	young generation
盲目	blindly
极	extremely
造成	cause
伤害	damage
认清	recognize
奴隶	slave

语法 Grammar

1. 却 (but/yet)

"却" is used as an adverbial modifier. It is placed in the second clause to indicate that the condition is not the same or is contrary to what was mentioned earlier.

Examples:

我以为我的答案是对的，结果却是错的。

I thought I gave the correct answer, but it turned out to be wrong.

身体健康是最重要的，很多人却刻意去减肥。

Health is the most important thing, but many people try so hard to lose weight.

2. 已经 (already)

This adverb indicates that something has already been done or is currently in progress.

Examples:

学校已经开学两个星期了。

School has already started for two weeks.

我的中文作业已经做完了。

I have already finished my Chinese homework.

3. 相反 (on the contrary/opposite)

"相反" can be used as an adverb at the beginning of a sentence. It indicates the opposite of what was mentioned earlier.

Examples:

很多人以瘦为美。相反，有些人觉得丰满的身形才是美。

Many people think thin equals beauty. On the contrary, some people think a plump body shape is beautiful.

她说的与我说的完全相反。

What she said is totally contrary to what I said.

4. 不只是 (more than/not only)

"不只是" can be placed before or after the subject to indicate that something is more than expected or something more is included.

Examples:

美不只是外在的容貌。

Beauty is more than outside appearance.

不只是我这样想，其他人也这样想。

I'm not the only one who thinks in this way.

词组及常用句式
Expressions & sentence structures

词组 Expressions

随处可见	omnipresent
翻开报纸	flip open the newspaper
对瘦的崇尚	obsession on keeping thin
成为一种潮流	become a trend
日渐改变	gradually changed
以瘦为美	regard thin as beautiful
唯一标准	the sole criterion for
衡量	to measure
其他的因素	other factors
天生容貌美丽	born beautiful
性格高傲	an arrogant personality
追求物质	materialistic pursuit
姣好的身材	an attractive figure
气质出众	refined
更多的是	what's more

过度追求	excessive pursuit of
导致不合理的价值观	lead to unreasonable values
盲目纤体减肥	blindly lose weight
极有可能	most likely
对他们的身体健康造成伤害	cause damage to their health

常用句式 Sentence structures

1. 以A (sth.) 为B (sth.)

Explanation	Example
consider A to be B	很多人以瘦为美。 Many people consider thinness to be beautiful.

2. 用A (sth.) 来衡量B (sth.)

Explanation	Example
use A to define B	我们不能用外表来衡量美。 We can't simply use appearance to define beauty.

3. A (sb./sth.) 对B (sb./sth.) 造成伤害

Explanation	Example
A causes damage to B	减肥会对我们的身体健康造成伤害。 Losing weight deliberately will cause damage to our health.

1.2 语言多样性
Language diversity

关键字 Key words

中文	英文	双语	重要	困难
语法	汉字	好处	兴趣	学习

模拟试题 Mock questions

SL	1. 你在学习中文的过程中遇到了哪些困难？你是怎么克服的？
	2. 请你谈谈学习中文的重要性。
	3. 你觉得中文老师应该怎么教中文？
	4. 请你谈谈学习中文的方法。
	5. 请你谈谈怎么保持语言水准。
HL	1. 请你谈谈双语学习的重要性。
	2. 请你介绍一下你所在城市的语言使用情况。
	3. 请你谈谈学习中文的经历和感受。
	4. 你觉得简体字和繁体字哪个更好？
	5. 请说说双语学习的困难和好处。

范文 Sample essay

你在学习中文的过程中遇到了哪些困难？你是怎么克服的？（SL）

学习中文的困难

Introduce the topic of learning Chinese.

我学习中文五年了。通过这五年的学习，我的体会是：对于西方人来说，中文很难学。下面我想谈谈我在学习中文的过程中遇到的困难。

Talk about your experience in learning Chinese.

总的说来，我很喜欢说中文。在我看来，中文语法不是很难，它是有规律的东西，很有意思，学起来也容易。虽然我的中文发音不是很标准，但是我喜欢和中国朋友练习对话。

Describe the difficulties you encountered when learning Chinese.

我在学习中文时遇到的最大困难就是写汉字了。汉字非常难记，一个读音有好几个汉字，有些字又有好几个读音。而且，汉字的书写很难，有些词我虽然知道怎么读，可是却写不出来，或者写成另外一个字。

Share your experience in how to overcome the difficulties.

中国有句话："兴趣是最好的老师"。意思就是说如果你对一件事情有兴趣的话，你就可以学好。所以我认为要想克服学习中文的困难，首先要对中文有兴趣。另外，要多听、多读、多写。我很喜欢学中文，所以我每天都会花至少两个小时学习中文。而且，我的一些中国朋友常常帮助我学习中文。因此，我相信，只要我继续努力，一定可以克服学习中文的困难。

(352 words)

词汇 Vocabulary

通过	through; by		兴趣	interest
体会	experience		如果	if
对于	for		所以	so
西方人	westerners		认为	think
学	learn		另外	in addition
下面	below		克服	overcome
谈谈	talk about		首先	firstly
过程	process		每天	every day
遇到	encounter		花	spend
困难	difficulty		至少	at least
语法	grammar		而且	moreover
规律	rule		一些	some
有意思	meaningful		常常	often
容易	easy		帮助	help
虽然	although		相信	believe
但是	however		继续	continue
标准	standard		努力	hardworking
练习	practice		一定	must
对话	conversation			
记	remember			
或者	or			

语法 Grammar

1. 总的说来 (generally/as a whole)

This phrase is used to describe a general situation or to make a general comment.

Examples:

总的说来，我是喜欢中文的。

Generally, I like learning Chinese.

总的说来，这件事情不能怪他。

As a whole, we can't blame him on this issue.

2. 可是 (but)

This conjunction is used to express the transition of the mood. It is similar to"但是"but has a weaker tone. Sometimes it can be used together with"却".

Examples:

我认得很多中文字，可是却写不出来。

I can recognize many Chinese characters, but I can't write them all out.

我不喜欢篮球，可是偶尔也会看篮球比赛。

I'm not really into basketball, but sometimes I will watch a basketball game.

3. 或者 (or)

This conjunction is used to introduce or connect different possibilities. It is different from"还是", which is only used in interrogative form.

Examples:

星期天我们会去爬山或者打篮球。

This Sunday we will go hiking or play basketball.

学生们可以学习简体字或者繁体字。

Students can learn simplified Chinese characters or traditional characters.

4. 至少 (at least)

The adverb "至少" indicates the minimum extent.

Examples:

她每天至少花两个小时学习中文。

She spends at least two hours studying Chinese every day.

这件衣服至少要五千块钱。

The shirt costs at least five thousand dollars.

词语及常用句式
Expressions & sentence structures

词组 Expressions

我的体会	my experience
对于西方人来说	for westerners
学习中文的过程中	the process of learning Chinese
总的说来	in general
在我看来	in my opinion
学起来也容易	easy to learn
不是很标准	not up to standard
和中国朋友练习对话	practice conversation with Chinese friends
汉字非常难记	Chinese characters are very difficult to remember
中国有句话	there is a saying in Chinese
对中文有兴趣	be interested in Chinese
至少两个小时	at least two hours
帮助我学习中文	help me to learn Chinese
继续努力	continue to work hard
克服学习中文的困难	overcome the difficulties of learning Chinese

常用句式 Sentence structures

1. 通过 (a period of time) 的学习

Explanation	Example
after a certain period of time	通过一年的学习，我的中文进步了。 After one year's study, my Chinese has improved.

2. 对于 (sb.) 来说

Explanation	Example
for somebody	对于西方人来说，中文很难学。 For westerners, learning Chinese is very difficult.

3. 在 (sb.) 看来

Explanation	Example
in somebody's point of view	在他看来，中文语法很简单。 In his point of view, Chinese grammar is easy.

4. 对 (sth.) 有兴趣

Explanation	Example
be interested in something	我对中文声调有兴趣。 I'm interested in Chinese intonations.

范文 Sample essay

请你谈谈双语学习的重要性。（HL）

双语学习的重要性

　　双语就是两种语言，很多学生都觉得学好英文就足够了，为什么还要多此一举学习中文呢？其实，学习两种语言对学生来说有很多好处。

Introduce the topic of bilingual study and state your viewpoint.

　　首先，学好两种语言可以帮你找到好工作。英文一直以来都是世界通用的语言，学好英语当然非常重要。但是近年来，中国的经济发展越来越快。如果你想到中国工作，就应该学好中文。即使不在中国工作，学习中文还是有助于我们找工作和提升竞争力，因为世界上四分之一的人会中文，说中文的人比说英语的人多。所以，懂中文的人就能够比那些不懂中文的人获得更多的工作机会。

Support your argument by talking about the first advantage of bilingual study.

　　其次，学会了中文可以了解更多博大精深的中国文化。我们可以上中文网站查阅资料，观看中文电影，听中文歌曲，这些都可以帮助我们更全面地了解中国文化。

Support your argument by talking about the second advantage of bilingual study.

　　再者，学两种语言可以令人更聪明。根据研究，学英语的人用的是左脑，而学中文用的是右脑；如果我们只学习一种语言，就只能训练一边的头脑，如果中英文一起学习，那么我们整个脑袋都能得到训练，我们就会变得更聪明伶俐。何况，小孩子和年轻人学习语言比较快，而且很少有口音。所以，我们在学校应该好好学习两种语言。

Support your argument by talking about the third advantage of bilingual study.

　　综上所述，我觉得双语学习对于中小学生很重要。

Conclude the essay by reaffirming your argument.

(436 words)

词汇 Vocabulary

双语	bilingual	机会	opportunity
语言	language	其次	secondly
觉得	think	了解	understand
足够	enough	查阅	look up
多此一举	superfluous	网站	website
其实	in fact	资料	information
好处	benefit	观看	watch
世界通用	universally used	电影	movie
当然	of course	歌曲	song
重要	important	全面	comprehensive
但是	but	博大精深	profound
近年来	in recent years	文化	culture
经济	economy	再者	again
发展	develop/development	聪明	smart
越来越快	more and more rapidly	根据	according to
如果	if	研究	research
工作	work	左脑	left brain
应该	should	右脑	right brain
即使	even if	训练	training
有助于	good for	一边	one side
提升	increase	一起	together
竞争力	competitiveness	那么	so
因为	because	整个	whole
懂	understand	脑袋	mind
更多	more	何况	let alone

小孩子	child		口音	accent
年轻人	young people		综上所述	to sum up

语法 Grammar

1. 即使……也／还是 (even if... still)

"即使" is used to say that in spite of what happens, the result is unaffected. It is usually used together with the adverbs "也" or "还是".

Examples:

> 即使明天下雨，我们还是要去看望张老师。
>
> Even if it rains tomorrow, we will still visit Ms Zhang.
>
> 即使考试考得很好，也不代表你所有的知识都会了。
>
> Even if you achieve a good result in the exam, it doesn't mean that you know everything.

2. ……分之……(fraction)

This expression indicates a fraction. The denominator is written first, then the numerator. The percentage expression is written as "百分之".

Examples:

> 世界上四分之一的人会说中文。
>
> Approximately one fourth of the world's population speaks Chinese.
>
> 百分之八十的人都看过这本书。
>
> Eighty percent of the people have read this book.

3. 何况 (let alone/besides)

This adverb is used to emphasize how unlikely or not possible a situation is because what was mentioned earlier is even as unlikely.

Examples:

> 这道题老师都不会做，何况学生。
>
> Even teachers cannot answer the question, let alone students.
>
> 我不会游泳，何况是潜水。
>
> I can't even swim, let alone dive.

4. 比 (compared with/than)

This preposition may be used to compare two or more things. It can be placed before an adjective or a verb to express comparison in a sentence.

Examples:

讲中文的人比讲英文的人多。

More people speak Chinese than English.

他的法语比我说得好。

He speaks French better than I do.

词组及常用句式
Expressions & sentence structures

词组 Expressions

对学生来说	for students
有很多好处	have many advantages
帮你找到好工作	help you find a good job
一直以来	always
世界通用的语言	universal language
当然非常重要	of course it is very important
近年来	in recent years
经济发展越来越快	the economy is developing more and more rapidly
有助于他们找工作	can help them find a job
世界上四分之一的人	one fourth of the world's population
有更多的机会	have a greater chance
了解更多中国文化	learn more about Chinese culture
查阅资讯	search for information
观看中文电影	watch Chinese movies
听中文歌曲	listen to Chinese songs

更全面地了解	understand something more thoroughly
令人更聪明	make people more intelligent
根据研究	according to studies
中英文一起学习	learn English and Chinese at the same time
得到训练	receive training
综上所述	in summary

常用句式 Sentence structures

1. 令／使 (sb.)

Explanation	Example
make/enable somebody do something or to be in a certain state	懂双语能**使**你找到好工作。 Knowing two languages may help you find a better job.

2. (sth./sb.) 越来越

Explanation	Example
more and more	中国的经济发展**越来越**快。 China's economy is developing more and more rapidly.

3. 综上所述，我觉得……

Explanation	Example
In summary, I think…	**综上所述，我觉得**我们应该从小就学习两种语言。 In summary, I think we should learn two languages starting from childhood.

4. 首先……，其次……，再者……

Explanation	Example
firstly,… secondly,… moreover,…	首先，学好双语可以使你找到好工作。其次，学习双语可以了解更多中国文化。再者，学双语能令人更聪明。 Firstly, acquiring two languages can help you find a good job. Secondly, it enables you to learn more about Chinese culture. Moreover, learning two languages makes people smarter.

Option topic 2
Customs and traditions
选修主题 2
风俗与传统

2.1 庆典、社会事件和宗教事件 Celebrations, social and religious events

关键字 Key words

节日	传统	庆祝	特色	重要
有名	气氛	丰富	热闹	重视

模拟试题 Mock questions

SL	1. 请你介绍一种中国传统节日的食物。
	2. 请谈谈中国人怎么庆祝传统节日。
	3. 你知道中国人是怎么过春节的吗？
	4. 请介绍一个你最喜欢的中国节日。
	5. 你生活的地区有什么特殊的节日庆祝活动？
HL	1. 你知道中国哪些传统节日的由来？
	2. 谈谈传统节日对于中国人的意义。
	3. 请介绍一个中国人重视的庆典活动。
	4. 请写一篇文章介绍中西方节日文化的不同。
	5. 现在，中国的很多年轻人更喜欢西方节日，忽视传统节日，你怎么看这种情况？

范文 **Sample essay**

请你介绍一下中国的传统节日。（SL）

中国的传统节日

中国的传统节日有很多，其中最有名的两个就是春节和中秋节了。

春节对于中国人来说是最重要的节日，因为这个时候家人不仅会一起过节，还会去亲友家里给他们拜年。春节的意义是告别旧的一年以及迎接新的一年。人们都希望，新的一年会有好的运气。春节的前一夜，中国人会和家人一起吃晚饭。那天的晚饭不但非常美味，而且会比平时的食物丰富。晚饭过后，人们也常常一起观看电视上的庆祝晚会，迎接新年。

中秋节也是一个很重要的节日。这一天，家人也会团聚在一起。中秋节的月亮特别圆，代表着团圆。所以，中秋节的时候，有些人虽然很忙，但也会回家和家人吃饭。通常晚饭过后，一家人会一边吃月饼一边赏月。很多小孩子晚饭后也会一起提着灯笼到街上或公园里玩耍，有时甚至玩到很晚。

从上面两个节日可以看出，中国人很重视家庭，团圆对于他们来说非常重要。所以，中国的传统节日都非常受人们喜爱。

Introduce two traditional Chinese festivals.

Talk about the meaning of the Spring Festival and how it is celebrated.

Talk about the meaning of the Mid-autumn Festival and how it is celebrated.

Give a general conclusion about Chinese festivals.

(343 words)

词汇 Vocabulary

传统	traditional	意义	significance/meaning	特别	especially
节日	festival	告别	bid farewell	圆	round
其中	among	旧	old	代表	represent
最有名	the most famous	以及	and	所以	so
春节	Spring Festival /Chinese New Year	迎接	welcome	忙	busy
中秋节	Mid-autumn Festival	希望	wish	通常	usually
		运气	luck	月饼	mooncake
对于	for	一起	together	赏月	moon gazing
最重要	the most important	晚饭	dinner	提着	carry
因为	because	非常	very	灯笼	lantern
家人	family	美味	delicious	街上	street
不仅	not only	而且	moreover	公园	park
过节	spend the holiday	平时	at ordinary times	玩耍	play
亲友	friends and relatives	食物	food	有时	sometimes
		丰富	a wealth of	甚至	even
拜年	visit people to wish them prosperity during the Spring Festival	观看	watch	重视	highly value
		电视	TV	家庭	family
		庆祝晚会	celebration party	喜爱	love
		月亮	the moon		

语法 Grammar

1. 其中 (among)

"其中" is usually placed at the beginning of the second clause. It cannot be placed after a noun.

Examples:

这里有各种各样的书，其中我最喜欢这本小说。

Among all these books, this novel is my favorite.

中国的传统节日有很多，其中最有名的是春节。

Among all the many traditional Chinese festivals, Spring Festival is the most well-known.

2. 然后 (then)

This construction is used to connect different actions or events and to indicate their order.

Examples:

我和姐姐在周末常常一起吃饭，然后看电影。

On weekends, my sister and I always eat together and then go see a movie.

你应该先做作业，然后再看电视。

You should do homework first. Then, you can watch TV.

3. 甚至 (even)

The adverb is used to add emphasis and is usually used with "也" or "都".

Examples:

他经常出去玩，有时候甚至玩到很晚。

He often goes out and sometimes even stays out until very late.

他每天都会踢足球，甚至下雨天也坚持。

He plays football every day, even on rainy days.

她学习很努力，甚至周日都在家学习。

She works very hard; she even studies at home on Sundays.

词组及常用句式
Expressions & sentence structures

词组 Expressions

传统节日	traditional festival
其中最有名的	among which the most famous is
对于中国人来说	to Chinese people
最重要的节日	the most important festival
给他们拜年	visit people to wish them prosperity during the Spring Festival
告别旧的一年	bid farewell to the end of the year
迎接新的一年	welcome the new year
春节的前一夜	the night before the Spring Festival
比平时的食物丰富	the food is more sumptuous than usual
晚饭过后	after dinner
特别圆	especially round
代表着团圆	represents reunion
一边吃月饼一边赏月	eating mooncakes while gazing at the moon
提着灯笼	carry lanterns
重视家庭	value the family
受人们喜爱	loved by many people

常用句式 Sentence structures

1. (activity/event) 过后

Explanation	Example
after a certain activity or event	晚饭过后，我们一起吃月饼。 After dinner, we ate mooncakes together.

2. (……) 的时候

Explanation	Example
when… takes place/during	中秋节的时候，家人都会团聚在一起。 During the Mid-autumn Festival, families will all gather together.

3. 从 (sth.) 可以看出

Explanation	Example
we can tell from something that	从春节可以看出，中国人很重视家庭。 From the Spring Festival, we can tell that Chinese people have very strong family values.

4. 受 (sb.) 喜爱

Explanation	Example
be loved/enjoyed by somebody	月饼很受中国人喜爱。 Mooncakes are a delicacy that many Chinese people enjoy having.

范文 Sample essay

你知道中国人是怎么过春节的吗？（HL）

中国的传统节日——春节

Introduce the topic of the essay: Spring Festival

春节是中国最重要的传统节日，它不仅是和家人团聚的日子，也是和亲戚朋友相互送祝福的佳节。

Talk about how people prepare for the festival and the festive cuisine.

临近春节时，人们会提前购买年货，比如春联和食物。贴春联是中国的传统习俗，春联上面通常写着寓意深刻的话，贴在门的两边代表着对新一年寄予的希望。至于春节的食物则五花八门，不同地方的风俗也不一样，比如北方人爱吃饺子，而南方人则喜欢吃年糕和鸡肉等。

Describe what people usually do on the day before the festival.

春节期间，除夕夜是最热闹的，因为除夕夜的时候全家人会团聚在一起吃年夜饭。除此之外，还会一起看电视上的春节晚会。接近凌晨的时候，人们还会一起倒数和放鞭炮来迎接新年。小孩子在除夕夜的时候特别开心，因为他们可以尽情地玩耍，甚至可以玩通宵。另外，他们还能从长辈那里收到很多红包。给红包是长辈对晚辈的一种祝福和爱护。

Further elaborate on the celebration activities.

春节的庆祝活动丰富多彩。大街上的每一个角落都热闹非凡，人们可以尽情享受节日的快乐。除了参加各式各样的庆祝活动，人们还会到亲戚朋友家拜年，互相说祝福的话，希望给自己和身边的人带来好运和吉祥。常见的祝福语有"恭喜发财"、"万事如意"和"心想事成"等。

Give a general conclusion about the festival.

春节期间中国到处洋溢着热闹愉快的节日气氛，所以大家都很喜欢过春节。

(438 words)

词汇 Vocabulary

亲戚	relative		放鞭炮	set off firecrackers
相互	mutual		尽情	enjoy
祝福	bless		通宵	stay up all night
佳节	festival		另外	in addition
临近	near		长辈	elders
提前	in advance		红包	red packet
购买	purchase		晚辈	the younger generation
年货	new year purchases for the Spring Festival		爱护	love and care
比如	for example		庆祝活动	celebration activity
贴春联	put up Spring Festival couplets		丰富多彩	a rich variety
习俗	custom		角落	corner
通常	usually		热闹非凡	lively
寓意深刻	profound meaning		享受	enjoy
寄予	wish		除了	except
希望	hope		参加	join
至于	to		拜年	visit people during the Spring Festival
年糕	rice cake		带来	bring
鸡肉	chicken		好运	good luck
除夕夜	New Year's Eve		吉祥	auspicious
接近	close to		常见	common
凌晨	the early hours of the morning		祝福语	greetings
倒数	count down		恭喜发财	wish someone wealth and prosperity

万事如意	good luck and fortune	到处	everywhere
心想事成	may all your wishes come true	洋溢	full of
		愉快	happy
期间	period	气氛	atmosphere

语法 Grammar

1. 除此之外 (apart from/other than)

This conjunction is used with "还" to express the idea that other than what has been mentioned previously, the following content is also included. When it is used with a negator, such as "不" and "没有", it expresses the notion that other than what has been mentioned before, nothing else is included.

Examples:

除夕夜家人会一起吃饭。除此之外，大家还会一起看庆祝晚会。

Family members will have dinner together on New Year's Eve. Apart from that, they will also watch the celebrations together.

她非常喜欢弹钢琴。除此之外，她没有其它爱好。

She likes playing piano very much. Other than that, she doesn't have any hobbies.

2. 另外 (in addition/what's more)

This conjunction connects two sentences. It expresses the idea that in addition to what has been mentioned in the previous sentence, the following content is also included.

Examples:

春节的时候，小孩子可以穿新衣服。另外，他们还可以从长辈那里收到很多红包。

During the Spring Festival, children can wear new clothes. What's more, they can get red packets from elder family members.

爸爸给我买了一台新电脑。另外，他还给弟弟买了新手机。

Dad bought me a new computer. What's more, he bought my brother a new cell phone.

3. 无论……都／也 (no matter how/what)

The phrase expresses the idea that something remains true whatever the situation is. In the first clause, the content expresses the notion "no matter what/how...". When the conjunction "还是" is used in the clause, it indicates an alternative.

Examples:

无论是大人还是小孩都很喜欢春节。

It doesn't matter whether you are an adult or a child— you will like the Spring Festival.

无论我怎么说，她都不相信。

No matter what I say, she won't believe it.

词组及常用句式
Expressions & sentence structures

词组 Expressions

最重要的传统节日	the most important traditional festival
相互送祝福	give blessings to each other
临近春节时	as the Spring Festival approaches
提前购买年货	buy New Year goods in advance
春联和食物	Spring Festival couplets and food
传统习俗	traditional custom
寓意深刻的话	words with profound meaning
贴在门的两边	attach on the sides of the door
寄予的希望	hope for
团聚在一起	get together
接近凌晨的时候	near midnight
迎接新年	welcome the New Year
特别开心	especially happy
玩通宵	play all night

长辈对晚辈的一种祝福和爱护	blessing and love of the elders to young people
春节期间	during the Spring Festival
每一个角落	every corner
尽情地享受	enjoy to the greatest extent
互相说祝福的话	say words of blessings to each other
带来好运和吉祥	bring good luck and good fortune
常见的祝福语	common words of blessing
到处洋溢	full of… everywhere
热闹愉快的节日气氛	lively and enjoyable festive atmosphere

常用句式 Sentence structures

1. 对 (sth./sb.) 的希望／祝福

Explanation	Example
hope/blessing for something or somebody	春联代表人们对新一年的希望。 Spring couplets reflect people's wishes for the new year.

2. (activity/event) 期间

Explanation	Example
during an event/activity	春节期间，人们会到亲戚朋友家拜年。 During the Spring Festival, people will visit their friends and relatives to wish them well in the New Year.

3. 给 (sb.) 带来 (sth.)

Explanation	Example
bring something to somebody	中国人相信祝福语会给别人带来好运。 Chinese people believe that saying words of blessings to others will bring them good luck.

2.2 中国传统饮食
Chinese traditional food and drinks

关键字 Key words

饮食	传统	特色	成为	习惯
食物	喜欢	特别	常见	包括

模拟试题 Mock questions

SL	1. 中国的传统食物有哪些？
	2. 谈谈你喜欢的中国传统食物。
	3. 中国传统饮食有什么特色？
	4. 你知道中国人的饮食习惯吗？
	5. 请介绍一下中国人吃饭时要注意的事情。
HL	1. 请你说说你所了解的中国饮食文化。
	2. 谈谈中国的茶文化。
	3. 请谈谈中国传统的饮食习惯。
	4. 请介绍中国的传统菜式。
	5. 请给你的朋友介绍一道中国的特色菜式。

范文 Sample essay

中国的传统饮食有哪些？（SL）

Introduce the topic by stating that Chinese cuisine has its own characteristics.

Talk about what people like to eat for breakfast in China.

Further talk about what people usually have for lunch and dinner in southern and northern China.

Give a general conclusion about Chinese cuisine.

中国传统饮食

中国的传统饮食很有特色，饮食习惯也和西方很不一样。

中国人常吃的早餐包括粥、包子、面条、鸡蛋等。南方的广东人早上喜欢去茶馆吃点心，最常见的点心包括虾饺、春卷等。很多人说的"饮茶"，其实就是早上去茶馆一边吃点心，一边喝茶。饮茶特别受老人欢迎，所以很多老人每天都会去饮茶。

至于午餐和晚餐，北方人喜欢吃面食，比如饺子、面条，但是南方人更爱吃米饭，很多人一日三餐都离不开米饭。而且，北方人比较经常吃肉，他们最喜欢羊肉和牛肉，而南方人则经常吃鱼和鸡肉。除了吃的食物不一样，南方人和北方人喜欢喝的东西也不同。北方人比较喜欢喝酒，而南方人比较喜欢喝汤。广东人更是每天都要喝汤，他们最喜欢喝鱼汤。但是，北方人和南方人都喜欢喝茶，比如绿茶和红茶。

虽然现在越来越多的中国人吃西餐，但是很多时候他们还是喜欢吃中餐，因为他们已经习惯了中国的传统饮食。

(330 words)

词汇 Vocabulary

传统	traditional	其实	in fact	牛肉	beef
饮食	diet	喝茶	drink tea	鱼	fish
特色	characteristic	特别	especially	鸡肉	chicken
习惯	habit	老人	the elderly	除了	except
西方	western	欢迎	popular among	不一样	different
早餐	breakfast	至于	as for	喝酒	drink wine
包括	include	午餐	lunch	喝汤	drink soup
粥	congee	晚餐	dinner	鱼汤	fish soup
包子	steamed bun	面食	food made from wheat	绿茶	green tea
面条	noodle			红茶	black tea
鸡蛋	egg	比如	for example	虽然	although
南方	southern	饺子	dumpling	现在	nowadays
茶馆	teahouse	但是	but	西餐	Western food
点心	dim sum	米饭	rice	中餐	Chinese food
常见	common	过去	in the past	因为	because
虾饺	shrimp dumpling	而且	moreover		
春卷	spring roll	经常	often		
等	and so on	吃肉	eat meat		
		羊肉	lamb		

语法 Grammar

1. 一边……一边…… (at the same time/while)

This construction indicates that two or more actions are taking place at the same time. It is placed before verbs.

Examples:

很多人喜欢早上去餐馆一边吃点心，一边喝茶。

A lot of people like to go to restaurants in the morning, eating dim sum while enjoying a cup of tea.

她一边弹琴，一边唱歌。

While she plays the piano, she sings at the same time.

2. 至于 (as for)

This conjunction indicates another aspect of what has been mentioned previously.

Examples:

至于晚餐，南方人喜欢吃米饭。

As for dinner, people in the south like having rice.

她每天都会去游泳。至于他，我就不知道了。

She goes swimming every day. As for him, I don't know.

3. 而 (while/but)

This conjunction has different meanings, one of which indicates that what is mentioned after is different from the content in the first clause. "而" is weaker than "但".

Examples:

北方人喜欢吃羊肉和牛肉，而南方人更喜欢吃鱼和鸡肉。

People in the north like lamb and beef, while people in the south prefer fish and chicken.

他的成绩很好，而我的成绩很差。

He got a high score, but I only got a low score.

词组及常用句式
Expressions & sentence structures

词组 Expressions

饮食习惯	eating habit
和西方很不一样	is very different from the West
常吃的早餐	common breakfast
去茶馆吃点心	have dim sum at tea houses
一边吃点心一边喝茶	have dim sum while drinking tea at the same time
受老人欢迎	welcomed by the elderly
至于午餐和晚餐	as for lunch and dinner
一日三餐	three meals a day
比较常吃肉	often eat meat
除了吃的食物不一样	apart from eating different kinds of food
比较喜欢	prefer
很多时候	most of the time
已经习惯了	become accustomed to

常用句式 Sentence structures

1. 和 (sth.) 不一样

Explanation	Example
different from something	中国的饮食习惯和西方不一样。 The eating habit of Chinese people is different from Western people.

2. 受 (sb.) 欢迎

Explanation	Example
be welcomed/liked by somebody (= popular)	点心很受年轻人欢迎。 Dim sums are very popular among young people.

3. 越来越多

Explanation	Example
more and more	越来越多的外国人喜欢吃中餐。 More and more foreigners like eating Chinese food.

范文 Sample essay

请你说说你了解的中国饮食文化。（HL）

中国四大菜系

中国菜在全世界都广受欢迎，世界各地开了很多中国饭馆。这说明中国的饮食文化已经传播到全世界。那么，中国菜究竟有哪些呢？

中国各个地方都有其独特的菜式，人们根据地域把中国菜分为四大菜系，分别是：四川菜、山东菜、江浙菜和广东菜。

四川菜又叫川菜，是中国最有特色的菜系。川菜之所以这么特别是因为它的味道丰富多样，包括麻辣和酸辣等，非常受人们欢迎。最著名的菜式有麻婆豆腐、宫保鸡丁和担担面。不仅中国人喜欢吃，很多外国人也赞不绝口。

山东菜又叫鲁菜，是中国另一个重要的菜系。它的特点是口味比较清淡。很多山东菜都用鱼来做，我们耳熟能详的鲁菜包括锅贴和糖醋鱼等。

江浙菜包括江苏菜和浙江菜。江浙菜讲究新鲜，经常用鱼和虾等海鲜来做原料，味道鲜美无比。而且江浙菜的烹饪方法是炖和蒸，食物既入味又清甜。其中一个特色菜是西湖醋鱼。

Introduce the topic by mentioning the popularity of Chinese cuisine.	
Briefly mention that Chinese cuisine is divided into four major types.	
Talk about Sichuan cuisine and some of its dishes.	
Talk about Shandong cuisine and some of its dishes.	
Talk about Jiangzhe cuisine and some of its dishes.	

Talk about Guangdong cuisine and some of its dishes.

—— 广东菜也叫粤菜，有很多不同的风味。粤菜使用的原料包括海鲜、鸡肉和牛肉等，味道也有很多变化，有些菜清淡，但有些菜香浓。点心是广东菜里面最有特色的，最受欢迎的点心有虾饺、春卷和烧卖等。

Give a general conclusion about Chinese cuisine.

—— 随着中国菜走向世界，全球很多人都对中餐情有独钟，中国的饮食文化也传播得越来越远。

(440 words)

词汇 Vocabulary

全世界／世界各地	all over the world
饭馆	restaurant
说明	explain
饮食文化	food culture
传播	spread
独特	unique
菜式	dishes
根据	according to
地域	region
菜系	cuisines
四川	Sichuan (province)
山东	Shandong (province)
江浙	Jiangsu and Zhejiang (provinces)
广东	Guangdong (province)
川菜	Sichuan cuisine
之所以	the reason for
味道	taste
丰富	plentiful
麻辣	spicy
酸辣	sour and spicy
最著名	the most famous
麻婆豆腐	spicy tofu
宫保鸡丁	spicy diced chicken with peanuts

担担面	Sichuan noodles with peppery sauce
不仅	not only
赞不绝口	full of praise
重要	important
口味	taste
比较	a little bit
清淡	light
耳熟能详	very familiar with something after having heard it for many times
锅贴	fried dumpling
糖醋鱼	sweet and sour fish
江浙菜	Jiangsu and Zhejiang cuisine
讲究	pay attention to
新鲜	fresh
海鲜	seafood
原料	raw ingredient
鲜美无比	incomparable freshness
烹饪方法	cooking method
炖	stew
蒸	steam
入味	marinate
清甜	fresh and sweet
西湖醋鱼	West Lake vinegar fish
粤菜	Guangdong cuisine

风味	flavor	虾饺	shrimp dumpling
原料	raw ingredients	春卷	spring roll
变化	change	烧卖	shao-mai (dumpling)
香浓	sweet and strong flavor	情有独钟	special preference for
点心	dim sum	传播	spread

语法 Grammar

1. 究竟 (what on earth)

This adverb is used in questions to emphasize the interrogative tone.

Examples:

中国菜究竟有哪些呢？

What on earth does Chinese cuisine contain?

你究竟怎么了？

What on earth happened to you?

2. 之所以……是因为 (the reason... is that)

This construction is used to emphasize the reason. "之所以" is placed after the subject in the result clause, while "是因为" is placed at the beginning of the second clause, indicating the reason.

Examples:

川菜之所以这么特别是因为它的味道丰富多样。

The reason Sichuan cuisine is so special is that it has rich flavors.

他成绩之所以这么好是因为他学习很努力。

The reason he gets high scores is that he works very hard.

3. 有些……有些 (some... while others)

"有些" functions as a modifier in this sentence structure. It refers to part of a collection of objects or a group of people.

Examples:

粤菜的口味有很多种，有些清淡，有些香浓。

There are different kinds of Cantonese cuisine. Some of them have a light flavor, while others have strong flavor.

他们去做运动了。有些人去打篮球，有些人去游泳。

They went to do exercises. Some are playing basketball, while others are going swimming.

词组及常用句式
Expressions & sentence structures

词组 Expressions

广受欢迎	very popular
世界各地	around the world
开了很多中国饭馆	a lot of Chinese restaurants have opened
传播到全世界	spread all over the world
各个地方	every place
独特的菜式	a unique dish
根据地域	according to regions
四大菜系	four major cuisines
之所以这么特别	the reason it is so special
丰富多样	plentiful
最著名的菜式	the most famous dish
口味比较清淡	the taste is a bit light
鱼和虾等海鲜	fish, shrimp and other seafood
味道鲜美无比	incomparable freshness
烹饪方法	cooking method

既入味又清甜	well-marinated and sweet
不同的风味	different flavors
走向世界	spread around the world
会传播得越来越远	will spread widely

常用句式 Sentence structures

1. 这说明

Explanation	Example
It shows that…	这说明中国的饮食文化已经传播到全世界。 It shows that the Chinese eating culture has already spread all over the world.

2. 把 (sth.) 分为

Explanation	Example
divide something into	人们把中国菜分为四个菜系。 Chinese cuisine can be divided into four major branches.

3. 包括A、B和C等

Explanation	Example
includes A, B and C, etc.	粤菜使用的原料包括海鲜、鸡肉和牛肉等。 The main ingredients of Guangdong cuisine include seafood, chicken and beef.

2.3 中国的礼仪礼节
Chinese etiquette

关键字 Key words

礼仪	礼节	注重	言行	习惯
修养	礼貌	遵守	风俗	体现

模拟试题 Mock questions

SL	1. 对于中国的礼仪，你知道多少？请谈谈你知道的情况。
	2. 你知道中国人的餐桌礼仪吗？
	3. 在日常生活中，中国有哪些礼仪？
	4. 请介绍一下中国的传统礼仪。
	5. 你的外国朋友要来中国，请提醒他要注意什么礼仪。
HL	1. 请你谈谈你所了解的中国传统礼仪。
	2. 谈谈礼仪和礼节在中国的重要性。
	3. 给你的外国朋友介绍一下中国的传统礼节。
	4. 中国的礼仪和礼节包括哪些方面？
	5. 在中国不同的场合里，人们应该注意哪些礼节？

范文 Sample essay

对于中国的礼仪，你知道多少？请谈谈你知道的情况。（SL）

中国的礼仪

Introduce the topic of Chinese etiquette.

中国人非常注重礼仪，在饮食、日常生活和人际交往方面都讲究礼仪。

Talk about table manners in Chinese culture.

首先，在饮食方面，请客吃饭就有很多礼仪。中国人请客总是准备丰盛的饭菜，可是主人往往还会对客人说"没做什么好吃的，请随便吃"之类的话，这些都是中国人礼貌的表现。在中国人家里吃饭，每当一道菜送上饭桌以后，主人总是会请客人先吃。每当这时，客人就会说些感谢的话或者请主人先吃。

Talk about Chinese etiquette in daily life.

其次，在日常生活方面，每个人的言行都体现着中国的礼仪。比如，在公共汽车上，如果看到老人和孕妇，中国人常常都会让座，因为这是对老人和孕妇的一种爱护和尊重。而平时，中国人也会特别照顾老人和小孩。

Talk about Chinese etiquette in interpersonal relationships.

另外，在人际交往方面，中国人也有很多礼节。如果中国人到朋友家里做客，会买一些水果或者食物送给朋友，而且一般都不会在朋友家待到很晚，因为那样会影响朋友休息。

Conclude the passage.

中国人一直都很注重礼仪，所以中国人一直被称为礼仪大国。

(334 words)

词汇 Vocabulary

中国人	Chinese people		或者	or
注重	place emphasis on		双方	both sides
礼仪	etiquette		其次	secondly
饮食	meal		言行	words and deeds
日常生活	daily life		体现	reflect
人际交往	interpersonal communication		比如	for example
首先	firstly		公共汽车	public bus
方面	aspect		老人	elderly
请客（吃饭）	host a meal for someone		孕妇	pregnant women
总是	always		常常	often
准备	prepare		因为	because
丰盛	great		爱护	love and care
饭菜	food		尊重	respect
可是	but		平时	normally
主人	host		特别	special
往往	often		照顾	care
客人	guest		另外	in addition
随便	casual		做客	visit someone
礼貌	polite		而且	furthermore
饭桌	dinner table		一般	general
以后	later		影响	influence
感谢	thank		休息	rest

语法 Grammar

1. 往往 (often/frequently)

This adverb indicates that a certain circumstance often happens.

Examples:

人们邀请她的时候，她往往拒绝。

When people invite her, she often declines.

这个季节往往会下大雨。

It often rains in this season.

2. 平时 (normally)

This adverb indicates what happens in normal circumstances.

Examples:

我平时很少去那家餐馆。

Normally, I seldom go to that restaurant.

她平时很懒惰，今天却干了两个小时家务。

Normally, she is very lazy, but she did housework for two hours today.

3. 一般 (generally)

The adverb indicates the activity that generally takes place.

Examples:

我一般很少去公园。

Generally, I seldom go to the park.

她一般每天看一个小时书，再看一个小时电视。

Every day, she generally reads for one hour and watches TV for another hour.

词组及常用句式
Expressions & sentence structures

词组 Expressions

注重礼仪	pay attention to etiquette
在饮食方面	in terms of meals
丰盛的饭菜	sumptuous dishes
请客人先吃	invite guests to start the meal first
每当这时	at this moment
感谢的话	words expressing thanks
在日常生活方面	in daily life
每个人的言行	each person's words and deeds
体现着中国的礼仪	embodies Chinese etiquette
爱护和尊重	love and respect
在人际交往方面	in terms of interpersonal relationships
到朋友家里做客	be a guest at a friend's home
待到很晚	stay until very late
影响朋友休息	disturb a friend's rest

常用句式 Sentence structures

1. 在 (sth.) 方面

Explanation	Example
when it comes to/in terms of something	在饮食方面，请客吃饭就有很多礼仪。 There are a lot of etiquette rules when it comes to hosting meals for guests.

2. 对 (sb.) 说

Explanation	Example
say something to somebody	主人会对客人说："请随便吃"。 The host will say "help yourselves" to the guests.

3. 每当……以后

Explanation	Example
after something happens	每当吃完饭以后，客人都会说"谢谢"。 After every meal, the guests will say "thank you."

范文 Sample essay

在中国不同的场合里，人们应该注意哪些礼节？（HL）

中国的礼节和礼仪

中国是个礼仪之邦，无论在什么场合，都非常讲究礼仪。因为懂礼仪既是显示自己修养的一种方式，也是尊重别人的一种表现。

首先，在一些正式的场合，比如参加会议的时候，要非常注意自己的言行举止。比如遇到了有较高社会地位的人，要用对方的职位来称呼，如"经理"和"董事长"等。与他人见面时可以握手，但不要拥抱，否则对方会觉得你过分热情。服装也要穿得很正式，因为得体的衣着是对别人的一种尊重。

其次，在一些特殊场合的时候，要遵守一些传统的风俗习惯。比如参加婚礼时，要穿鲜艳的服装，最好是红色。因为红色对中国人来说是吉利的颜色，他们认为在喜庆的日子里穿红色的衣服，会带来好运。另外，客人也要给适当的礼金，表示对新人的祝福。

在其他场合，中国人也要遵守一定的礼仪。比如吃饭的时候不能用筷子指着别人，也不能把筷子插在饭里，更不能用筷子敲打碗碟。

中国人相信，只要人人懂礼仪、遵守礼仪，社会就会更和谐，人们的生活也会更美好。

Introduce the topic by stating that China has often been referred to as the Nation of Etiquette.

Explain how people should behave and address people in formal occasions.

Explain how people should behave in special occasions, such as weddings.

Further elaborate on other etiquette rules.

Conclude by stressing the importance of practicing good etiquette.

(361 words)

词汇 Vocabulary

无论	no matter		服装	clothes
场合	occasion		衣着	clothing
讲究	pay attention to		特殊	special
修养	upbringing/ refinement		遵守	comply with
方式	way		传统	traditional
表现	performance		风俗习惯	custom
正式	formal		婚礼	wedding
参加	participate in		鲜艳	bright
会议	meeting		吉利	lucky
举止	behavior		认为	think
较高	higher		适当	appropriate
社会地位	social status		礼金	monetary gift
职位	position		新人	newlyweds
称呼	call/ address		祝福	blessing
经理	manager		相信	believe
董事长	director		人人	everyone
握手	shake hands		社会	society
拥抱	hug		和谐	harmonious
否则	otherwise		生活	life
过分	excessive		美好	good
热情	enthusiasm			

语法 Grammar

1. 否则 (otherwise)

This conjunction expresses the idea that a certain result will arise if the conditions mentioned earlier are not met.

Examples:

> 在图书馆时我们要小声说话，否则会打扰到别人。
>
> We should speak quietly in the library. Otherwise, we will disturb other people.
>
> 你必须参加这次晚会，否则，他是不会参加的。
>
> You have to attend the party. Otherwise, he won't attend either.

2. 最好 (had better/it's the best)

"最好" indicates the most ideal situation or the best option for somebody.

Examples:

> 明天最好不下雨，这样我们才能出去玩。
>
> It would be best if it didn't rain tomorrow; then we could we go out.
>
> 你最好早点出发，否则可能迟到。
>
> You'd better leave earlier; otherwise, you may be late.

3. 只要……就 (as long as... then)

This construction indicates that a certain result will occur once a condition is met. "只要" can be placed before or after the subject in the condition clause, while "就" introduces the result.

Examples:

> 只要你懂得怎么处理，这件事情就不会有负面影响。
>
> As long as you know how to deal with it, there won't be any negative influence.
>
> 你只要听医生的话，就能很快康复。
>
> As long as you follow the doctor's instructions, you will recover soon.

词组及常用句式
Expressions & sentence structures

词组 Expressions

礼仪之邦	nation of etiquette
无论在什么场合	no matter what occasion it is
懂礼仪	know about etiquette rules
显示自己修养的一种方式	a way of showing your refinement
尊重别人的一种表现	a way of showing respect for others
正式的场合	a formal occasion
参加会议	attend a meeting
言行举止	how you speak and act
较高社会地位的人	people of higher social status
对方的职位	the position of others
称呼对方	address others
过分热情	overly enthusiastic
穿得很正式	formally dressed
得体的衣着	respectable clothing
传统的风俗习惯	traditional custom
参加婚礼	attend a wedding
鲜艳的服装	brightly colored clothing
最好是红色	red is the best
吉利的颜色	auspicious color
在喜庆的日子里	on festive days
适当的礼金	appropriate gift value
指着别人	point at other people with your finger
敲打碗碟	make clanking noises with the bowls and plates
遵守礼仪	practice good etiquette

常用句式 Sentence structures

1. ……的一种方式

Explanation	Example
a way of (doing something)	懂礼仪是表现自己修养的一种方式。 Practicing good etiquette is a way of showing one's refinement.

2. ……的一种表现

Explanation	Example
a manifestation of something	这是尊重别人的一种表现。 This is a manifestation of showing respect to others.

3. 对 (sth./sb.) 的尊重

Explanation	Example
respect for somebody or something	在正式的场合，穿着正式是对别人的尊重。 In formal occasions, dressing up is a way of showing respect for others.

Option topic 3
Health

选修主题 3
健康

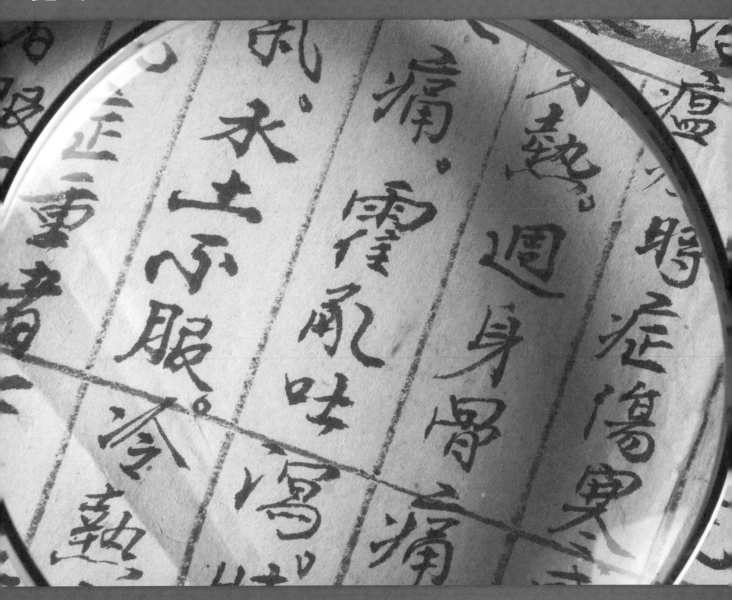

3.1 保健服务
Health services

关键字 Key words

健康	保健	身体	生活	注意
饮食	运动	检查	心态	身心

模拟试题 Mock questions

SL	1. 请你说说怎样才能保持健康的身体。
	2. 青少年应该如何处理压力来保持身心健康？
	3. 请介绍一种健康食物。
	4. 请谈谈应该如何看待压力。
	5. 什么才是健康的生活方式？
HL	1. 请你谈谈怎样拥有健康的身心。
	2. 谈谈健康和饮食的关系。
	3. 谈谈健康和运动的关系。
	4. 请谈谈什么是健康的饮食习惯。
	5. 请给你的朋友介绍健康的生活方式。

范文 Sample essay

请你说说怎样才能保持健康的身体。（SL）

如何保持健康的身体

随着我们的生活条件越来越好，我们能享受的东西也越来越多。但是如果没有健康的身体，一切就没有了意义。因此，为了能够享受更好的生活，我们应该注意身体健康。要过健康的生活，就要从下面几个方面努力：

第一，要注意自己的饮食。一日三餐是每天必须的，有些人不吃早餐，其实这样会损害我们的身体健康。我们不但要吃足三餐，而且要吃得好。多吃水果、蔬菜、米饭和面条，少吃垃圾食品，如薯片、饼干等。

第二，要早睡早起。早睡早起对身体好，但是很多年轻人晚上很晚睡觉，第二天很晚起床，也不吃早餐，身体因此受到影响。

第三，要多运动。多运动身体才会更强壮，还可以减少疾病。我们有时间应该多参加运动，锻炼好身体。

第四，要每年检查身体。很多疾病如果早发现就能早点治疗，因此我们要注意身体，如果感觉不舒服就要尽快去检查。

保持一个健康的身体对每个人都非常重要。我们要从现在开始为我们的健康而努力。

Explain the importance of being healthy.

Introduce four ways to stay healthy: balanced diet, adequate rest and sleep, more exercise and regular body checks.

Summarize the passage.

(344 words)

词汇 Vocabulary

随着	with	每天	every day	起床	get up		
条件	condition	必须	must	影响	influence		
享受	enjoy	有些人	some people	运动	sport		
但是	but	早餐	breakfast	强壮	strong		
如果	if	其实	in fact	减少	reduce		
健康	health	损害	damage	疾病	disease		
身体	body	不但	not only	时间	time		
意义	meaning	水果	fruit	参加	take part in		
因此	so	蔬菜	vegetables	锻炼	exercise		
为了	in order to	米饭	rice	检查	check		
更好	better	面条	noodles	发现	find		
生活	life	垃圾食品	junk food	治疗	treatment		
应该	should	薯片	potato chips	感觉	feel		
注意	pay attention to	饼干	cookies	不舒服	uncomfortable		
方面	aspect	年轻人	young people	拥有	have		
努力	effort	晚上	night	重要	important		
饮食	diet	睡觉	sleep	开始	begin to		

语法 Grammar

1. 随着 (with)

This preposition is placed at the beginning of the first clause to indicate the influence of a certain condition, followed by the result in the second clause.

Examples:

随着生活条件的提高，我们能享受更多的便利。

With the improvement of living conditions, we can enjoy more convenience.

随着天气的变化，越来越多人患上感冒。

With the changing weather, more and more people may catch a cold.

2. 其实 (in fact)

This adverb indicates that the following content is true, supplementing what has been said before.

Examples:

我以为法语很难，其实不是的。

I thought learning French would be difficult, but in fact, it isn't.

这个问题看起来很复杂，其实很简单。

The problem looks complicated, but in fact, it is quite simple.

3. 不但……而且 (not only... but also)

This construction indicates a progressive relationship. It is similar to "不仅……还……". If the two clauses share the same subject, the subject is placed in front of "不但" in the first clause. If the two clauses have different subjects, "不但" and "而且" are placed before their respective subjects.

Examples:

我们不但要一日吃足三餐，而且要吃得好。

We should not only eat three meals per day, but also eat well.

不但他不相信，而且其他人也不相信。

Not only does he not believe it, neither do others.

4. ……就…… (then... right away)

This adverb suggests that a certain action takes place right after the previous one. Thus, "就" is often placed between two verbs or verbal phrases.

Examples:

很多疾病早发现就能早治疗。

If found early, many diseases can be treated right away.

她到了家就做功课。

She does homework right after she arrives home.

词组及常用句式
Expressions & sentence structures

词组 Expressions

生活条件	living condition
越来越好	getting better and better
没有了意义	meaningless
享受更好的生活	enjoy a better life
注意身体健康	pay attention to health
过健康的生活	live a healthy life
从下面几个方面努力	make efforts in the following aspects
注意自己的饮食	pay attention to one's diet
一日三餐	three meals a day
损害我们的身体健康	damage our health
吃足三餐	eat three meals per day
垃圾食品	junk food
早睡早起	early to bed and early to rise
对身体好	good for the body
很晚睡觉	sleep late
很晚起床	get up late
身体因此受到影响	thus health is affected
多运动	exercise more
减少疾病的发生	reduce the occurrence of disease
参加运动	participate in sports activities
锻炼好身体	exercise to maintain good health
检查身体	to check the body
感觉不舒服	feel uncomfortable

尽快去检查	have a body check as soon as possible
保持一个健康的身体	keep a healthy body
对每个人都非常重要	very important for everyone
从现在开始	from now on
为我们的健康努力	make an effort to improve our health

常用句式 Sentence structures

1. 对 (sb./sth.) 好

Explanation	Example
benefit somebody or something	经常运动对身体好。 Doing sports frequently benefits our health.

2. (sb./sth.) 受到影响

Explanation	Example
something is affected	不吃早餐，身体会受到影响。 Without breakfast, our health will be affected.

3. 从 (time) 开始

Explanation	Example
start from a certain time	从现在开始我们要注意身体。 From now on, we should pay attention to health.

范文 Sample essay

青少年应该如何处理压力来保持身心健康？（SL）

<table>
<tr><td>

Explain the importance of dealing with pressure properly.

</td><td>

正确处理压力

　　经常有学生因不懂如何处理压力而自杀，所以，学习减压是非常重要的。学生的压力是哪里来的呢？我们应该怎样正确处理呢？

</td></tr>
<tr><td>

Analyze the main sources of pressure: school and parents.

</td><td>

　　学生的压力主要来源于学校和家庭。在学校里，同学之间的竞争很大。大家都非常认真，但是第一名只有一个。为了争取好成绩，我们不得不放弃很多休息或者玩的时间去读书。放学回到家，我们的压力不仅没有减少，反而会变得更大。很多父母对孩子的要求很严格，期望也很高。我们不想让父母失望，所以压力越来越大。

</td></tr>
<tr><td>

Explain that too much pressure will cause negative effects and suggest two ways to reduce pressure.

</td><td>

　　有人说压力可以产生动力，我十分赞成。然而，压力不可以过度，否则会起到相反的作用。青少年应该学会自我调整，这样才能克服压力。减压最好的方法是跟别人谈谈你为什么有压力，这样有助于我们把不安抒发出来。此外，旅行也是减压的好方法。旅行的时候，你会觉得世界很大，而你所面对的问题其实没什么，没有必要那么烦恼。

</td></tr>
<tr><td>

Summarize the passage.

</td><td>

　　青少年应该正确处理压力，让压力变为我们进步的动力。

</td></tr>
</table>

(344 words)

词汇 Vocabulary

青少年	teenager		严格	strict
如何	how		期望	expect
压力	stress		失望	disappoint
保持	keep/maintain		产生	produce
经常	often		动力	motivation
自杀	suicide		赞成	agree
减压	reduce stress		然而	however
主要	main		过度	excessive
来源	source		否则	otherwise
学校	school		相反	in contrast
家庭	family		作用	role
竞争	competition		调整	adjust
大家	everyone		克服	overcome
认真	seriously		谈谈	talk about
争取	strive for		有助	helpful
成绩	grades		不安	uneasy
不得不	have to		抒发	express
放弃	give up		此外	in addition
休息	rest		旅行	travel
或者	or		方法	method
放学	after school		必要	necessary
反而	instead		烦恼	worry
父母	parents		转化	transform
孩子	child		进步	improve
要求	expectation			

语法 Grammar

1. 因……而…… (due to)

This construction shows a causal relationship. "因" introduces the reason, while "而" introduces the corresponding result.

Examples:

有些学生因不懂如何处理压力而自杀。

Some students commit suicide due to the fact that they don't know how to deal with pressure.

这名运动员因不小心而受了伤。

The athlete was injured due to carelessness.

2. 不得不 (have to)

This construction indicates that a certain action must be performed due to some condition or in order to achieve some result.

Examples:

为了争取好成绩，她不得不放弃周末休息的时间。

In order to achieve good results, she has to give up her rest on weekends.

马上要下雨了，我们不得不赶紧回家。

It's going to rain any minute. We have to go home right away.

3. 然而 (however)

This conjunction serves as the transition in the sentence. It is placed at the beginning of a sentence or clause.

Examples:

有人说压力可以产生动力。然而，压力不可以过度。

Some people believe that pressure can serve as a driving force. However, pressure cannot exceed a certain limit.

她说的可能是真的，然而我们不能全信。

What she says might be true. However, we should not trust her completely.

4. 此外 (in addition)

This conjunction indicates that on top of what has been mentioned, the following content is also included.

Examples:

听音乐能减压，此外，旅行也可以减压。

Listening to music can reduce pressure. In addition, travel can also help relieve stress.

她会弹吉他、拉小提琴，此外还会弹钢琴。

She can play the guitar and violin. In addition, she also plays the piano.

词组及常用句式
Expressions & sentence structures

词组 Expressions

主要来源于学校和家庭	comes mainly from school and family
同学之间的竞争很大	fierce competition among students
为了争取好成绩	in order to achieve good results
不得不放弃	have to give up
对孩子的要求很严格	be strict with children
期望很高	high expectation
不想让父母失望	do not want to let parents down
压力可以产生动力	pressure can serve as a driving force
不可以过度	cannot exceed a certain limit
起到相反的作用	lead to the opposite effect
学会自我调整	learn to adjust oneself
克服压力	overcome pressure
跟别人谈谈你为什么有压力	talk with others about the pressure you face
面对的问题	the problems faced by

| 没有必要那么烦恼 | there is no need to be so worried |
| 进步的动力 | driving force to improve |

常用句式 Sentence structures

1. 来源于 (sth./somewhere)

Explanation	Example
come from something/ somewhere	压力来源于学校和家庭。 Pressure comes from schools and families.

2. 对 (sb.) 的要求很严格

Explanation	Example
have high expectations of	父母对孩子的要求很严格。 Parents have high expectations of their children.

3. 起到 (adj.) 作用

Explanation	Example
lead to some effect	压力过度会起到相反的作用。 Excessive pressure will lead to the opposite effect.

范文 Sample essay

请你谈谈怎样拥有健康的身心。（HL）

如何拥有健康的身心

大家常说："健康是福"，这话一点儿也不假。只有拥有健康的身体，我们才能正常地学习、工作和生活。

很多人整日忙碌，在工作上取得很大的成就，但常常忽视身体的重要性。熬坏了身体，再大的成就也没法享受。因此，为了自己的身体健康，我们要从下面几个方面好好照顾自己。

第一，注意饮食。除了要按时吃饭，还要注意饮食的营养。多吃蔬菜、水果，但不要只吃同一种蔬菜，而要吃不同种类的。另外，少吃肉和蛋等食物。

第二，正常作息。很多年轻人会熬夜玩游戏，第二天又早起去上课，因为睡眠时间不够而影响了身体健康。我们应该改掉这种不健康的作息方式，早睡早起。

第三，经常运动。长时间学习、工作，身体很容易疲劳，运动能够缓解疲劳，使人放松心情，有益于身体健康。因此，我们有时间就要经常做运动，锻炼身体。

第四，检查身体。人一旦生病就要很长时间才能痊愈，如果能定期体检，就能及时发现问题，避免疾病的发生。

第五，调整心态。每个人都会遇到不开心的事情，当不好的事情发生时，要调整自己的心态，不要让负面的情绪影响自己的身体。

拥有健康的身心，才能更好地享受生活。大家应该记住"健康是福"，从各方面保持身心健康。

Introduce the topic by explaining the importance of health.

This serves as a transitional paragraph leading to different ways to stay healthy.

List five different aspects that we should pay attention to in order to stay healthy.

Summarize the passage by stressing the importance of staying healthy.

(426 words)

词汇 Vocabulary

正常	normal	应该	should
整日	all day	运动	do sports
忙碌	busy	长时间	for a long time
取得	obtain	容易	easy
成就	achievement	疲劳	fatigue
常常	often	缓解	ease
忽视	ignore	放松	relax
重要性	importance	心情	mood
没法	can't	有益于	beneficial to
照顾	take care of	一旦	once
除了	except	痊愈	recover
按时	on time	定期	regular
营养	nutrition	体检	body check
种类	type	发现	find
另外	in addition	避免	avoid
肉	meat	发生	happen
熬夜	stay up late	心态	attitude
睡眠	sleep	遇到	encounter
不够	not enough	情绪	mood
影响	influence		

语法 Grammar

1. 为了 (in order to)

This phrase is used to express a purpose. It can be placed before or after the subject in the first clause.

Examples:

为了身体的健康，我们要平衡作息。

In order to stay healthy, we should balance work and rest.

妈妈为了照顾家庭，每天都早起晚睡。

In order to take care of the family, my mother gets up early and goes to bed late every day.

2. 一旦……就 (once)

This construction expresses the idea that a certain result will likely occur under some circumstance.

Examples:

人一旦生病就要很长时间才能康复。

Once people get sick, it may take a long time for them to recover.

一旦发现问题，就要马上解决。

Once you discover a problem, you should solve it immediately.

3. 当……时 (when)

This structure indicates the time that an action or event takes place. It usually forms the first clause, followed by the second clause, which starts with the subject.

Examples:

当不好的事情发生时，我们要调整好自己的心态。

When something bad happens, we should adjust our minds and attitudes.

当知道他赢了比赛时，他妈妈高兴得不得了。

When his mother heard that he had won the competition, she was overjoyed.

词组及常用句式
Expressions & sentence structures

词组 Expressions

大家常说	people often say
一点儿也不假	cannot be more true
拥有健康的身体	have good health
整日忙碌	busy all day
在工作上取得很大的成就	achieve great success at work
忽视身体的重要性	ignore the importance of health
再大的成就	in spite of great achievement
没法享受	cannot enjoy
为了自己的身体健康	for their own health
好好照顾自己	take good care of oneself
注意饮食	be mindful of what we eat
按时吃饭	eat on time
不同种类	different types
正常作息	fixed daily routine
熬夜玩游戏	stay up all night playing games
影响了身体健康	influence their health
早睡早起	early to bed and early to rise
保持一个健康的身体	maintain a healthy body
经常运动	exercise often
缓解疲劳	relieve fatigue
使人放松	make people relax
有益于身体健康	beneficial to health
定期体检	get a body check regularly
及时发现问题	find out problems in time

检查身体	body check
避免疾病的发生	avoid disease
调整心态	adjust one's attitude
遇到不开心的事情	encounter unhappiness
享受生活	enjoy life
大家应该记住	everyone should remember

常用句式 Sentence structures

1. 一点儿也不 (adj.)

Explanation	Example
cannot be more/ not... at all	这句话一点儿也不假。 These words cannot be more true.

2. (sth.) 的重要性

Explanation	Example
the importance of something	我们不能忽略健康的重要性。 We cannot ignore the importance of health.

3. 在 (sth.) 上取得成就

Explanation	Example
achieve success in something	有些人虽然在工作上取得很大的成就，但身体健康受到影响。 Some people sacrifice their health in pursuit of success at work.

4. 避免 (sth.) 的发生

Explanation	Example
prevent something from happening	我们要注意身体，避免疾病的发生。 We should pay attention to our own bodies in order to prevent diseases.

3.2 流行病
Epidemic disease

关键字 Key words

感冒	咳嗽	疼痛	严重	预防
防止	生病	日常	作息	锻炼

模拟试题 Mock questions

SL	1. 感冒是常见的疾病，请谈谈应该如何预防感冒。
	2. 你生过病吗？请谈谈生病的感受。
	3. 讲述一次生病的起因和经历。
	4. 你平时是怎么预防感冒的？
	5. 请说说常见的疾病有哪些。
HL	1. 春天是疾病高发期，你知道哪些预防措施？
	2. 请介绍一种常见的流行病及其预防措施。
	3. 如何预防常见的流行性疾病？
	4. 请讲述一次生病住院的经历，并说说你的感受。
	5. 你的朋友不小心生病了，请告诉他要注意哪些方面。

范文 Sample essay

感冒是常见的疾病，请谈谈应该如何预防感冒。（SL）

如何预防感冒

感冒是常见的疾病，发病时可能会伴有咳嗽、流鼻涕、喉咙疼痛，食欲减退等症状，如果严重的话，还会导致发烧和头痛。因此，我们要提前预防，以免生病。

要预防感冒，可以从下面几个方面做起：

一、适时增减衣服。春天刚过时，不要马上减衣，气温下降时，要及时多穿衣服。尤其是夏冬季节室内外温差较大时，更要注意防寒保暖。

二、注意卫生。要经常保持室内清洁，使室内空气流通，阳光充足。吃饭前和上厕所之后要洗手，如果从外面回到家里也要洗手。另外，避免与感冒患者接触，在感冒多发期应带口罩，尽量少在公共场合出入。

三、均衡饮食。日常饮食中，不仅要吃适量鱼、肉、鸡、蛋，还应多吃蔬菜和水果以补充维生素。另外，少吃零食和垃圾食品，多喝温开水。

四、保证充足睡眠。过度疲劳会影响健康。有充足的睡眠，会增加身体的抵抗力，这样才能更好地预防感冒。

五、多做室外活动。要多到室外活动，比如跑步、打篮球、爬山等。多晒太阳、多呼吸新鲜空气能使身体更强壮，不易受感冒病毒的侵袭。

如果在日常生活中能做到以上几点措施，就能预防感冒。

Introduce the common cold and its symptoms.

List five aspects that we should pay attention to in order to avoid catching a cold.

Conclude the passage by stressing the importance of disease prevention.

(384 words)

词汇 Vocabulary

感冒	cold	衣服	clothes	不仅	not only
常见	common	春天	spring	适量	right amount
疾病	disease	马上	immediately	鱼	fish
如何	how	减衣	reduce clothing	肉	meat
预防	prevent/ prevention	气温	temperature	鸡	chicken
可能	may	下降	drop	蛋	egg
伴有	with	及时	in time	蔬菜	vegetables
咳嗽	cough	注意	pay attention to	另外	in addition
喉咙	throat	卫生	hygiene	零食	snacks
疼痛	pain	经常	often	垃圾食品	junk food
严重	serious	保持	keep	保证	ensure
导致	lead to	室内	indoor	睡眠	sleep
发烧	fever	清洁	clean	发生	happen
头痛	headache	空气	air	室外活动	outdoor activities
因此	so	新鲜	fresh	爬山	mountain climbing
提前	in advance	阳光	sunshine	呼吸	breathe
生病	sick	充足	enough	强壮	strong
下面	below	厕所	toilet	如果	if
方面	aspect	均衡	balanced		
适时	timely	饮食	diet		
		日常	daily		

语法 Grammar

1. 因此／所以 (therefore/so)

This conjunction indicates the result or conclusion. "因此" can be placed before or after the subject in a clause, while "所以" can only be used before the subject.

Examples:

天气变冷了，因此我们要多穿点衣服。

The weather is becoming cold. Therefore, we should put on more clothes.

这场比赛很重要，所以我们不能放弃。

This game is really important, so we should not give up.

2. 既……又 (also/and)

This construction indicates two concurrent situations or qualities.

Examples:

好的睡眠既有利于身体，又增强抵抗力。

Quality sleep is good for one's health. It also enhances one's immune system.

这个学生既聪明又努力。

The student is smart and hardworking.

3. 如果……就 (if... then)

This construction is used to show a conditional relationship. "如果" in the first clause introduces the condition, while "就" in the second clause indicates the result or conclusion brought about by the condition.

Examples:

如果明天天气不好，我们就不去爬山了。

If the weather is not good tomorrow, then we will not go mountain climbing.

如果我们能克服困难，就能取得成功。

If we can overcome the difficulties, then we can achieve success.

词组及常用句式
Expressions & sentence structures

词组 Expressions

最常见的疾病	the most common diseases
伴有咳嗽、喉咙疼痛	accompanied by cough and sore throat
导致发烧和头痛	leading to fever and headache
提前预防	prevent in advance
以免生病	prevent illness
（更好地）预防感冒	prevent catching a cold (more effectively)
从下面几个方面做	take action in the following aspects
适时增减衣服	change the clothes you wear in a timely manner
马上减衣	wear fewer clothes at once
气温下降时	when temperature drops
注意卫生	pay attention to hygiene
经常保持室内清洁	keep the indoor environment clean
使室内空气新鲜	keep the air fresh indoors
阳光充足	sunny
均衡饮食	balanced diet
多吃些蔬菜和水果	eat more vegetables and fruits
保证充足睡眠	ensure adequate sleep
多做室外活动	do more outdoor activities
多晒太阳	sunbathe more
多呼吸新鲜空气	get more fresh air
使身体更强壮	make the body stronger

常用句式 Sentence structures

1. 最常见的 (sth.)

Explanation	Example
the most common	感冒是最常见的疾病。 A cold is one of the most common diseases.

2. 导致 (sth.)

Explanation	Example
lead to a certain result	感冒很容易导致发烧和头痛。 Having a cold may easily lead to fever and headache.

3. 注意 (sth.)

Explanation	Example
pay attention to something	请注意，最后一班列车将于十一时开出。 Please note that the last train will depart at 11 p.m.

范文 Sample essay

春天是疾病高发期，你知道哪些预防措施？（HL）

春季疾病的预防

Explain that many diseases often occur in spring.

春季是各种疾病的多发季节，如果不提前做好预防措施，就有可能受到它们的侵扰。

Introduce the common cold and its symptoms.

普通流行性感冒是一种常见的春季传染病，简称"流感"。其中，流感病人为传染源，主要在人多拥挤的环境中经空气传播。患病时会出现发热、头痛、乏力和咳嗽等症状，还可能出现肠胃不适。

Introduce allergy and its symptoms.

而最常见的春季非传染病即是人们常说的"过敏"。春天风大，空气中有很多花粉、柳絮等导致过敏的因素。过敏的人皮肤会呈现干燥红肿现象，而且非常痒。

提前预防各种疾病的措施主要有：

List different measures we should take to prevent diseases.

一、学习、了解卫生知识，树立防病意识。

二、接种相应的疫苗，是抵抗传染病发生的最佳手段。

三、注意调整作息、合理锻炼，增强抵抗疾病的能力。

四、开窗通气，少去空气不流通、人多拥挤的场所，尤其是体弱者。

五、注意个人卫生，养成良好的卫生习惯，饭前便后要勤洗手。

六、遇到气候变化，注意增减衣服。

七、有过敏体质的人应尽量少去赏花和接触花粉，尽量避免风吹日晒。

八、外出时要戴口罩。

如果一旦感染流感，就要及时吃药并接受治疗。另外，万一过敏发病，不要吃刺激性食物，如辣椒、生葱、生蒜，必要时到医院就医。

Summarize the passage.

(396 words)

词汇 Vocabulary

措施	measure	肠胃	gastrointestinal
春季	spring	不适	discomfort
疾病	disease	过敏	allergic
季节	season	花粉	pollen
侵扰	invade	柳絮	catkin
传染病	infectious disease	导致	lead to
普通	ordinary	因素	factor
流行性	epidemic	皮肤	skin
感冒	cold	呈现	appear
简称	abbreviation	干燥	dry
其中	among them	红肿	red and swollen
流感病人	influenza patient	现象	phenomenon
传染源	source of infection	而且	moreover
主要	main	痒	itchy
拥挤	crowded	了解	understand
环境	environment	知识	knowledge
空气	air	树立	set up
传播	spread	防病	prevent disease
患病	sick	意识	aware of/awareness
出现	appear	接种	vaccinate/vaccination
发热	fever	相应	corresponding
头痛	headache	疫苗	vaccine
乏力	weak	抵抗	resist
咳嗽	cough	最佳	the best
症状	symptom	手段	means

注意	pay attention to		外出	go out
合理	reasonable		戴	wear
锻炼	exercise		口罩	mask
能力	ability		如果	if
开窗	open the window		一旦	once
通气	ventilate		感染	infect/infection
不流通	poor circulation		及时	in time
场所	place		接受	receive
尤其	especially		治疗	treatment
体弱者	people with poor health		另外	in addition
个人	personal		万一	in case
养成	develop		发病	develop symptoms of a disease
习惯	habit		刺激性	irritating
遇到	encounter		辣椒	chili
气候	climate		生葱	onion
变化	change		生蒜	garlic
尽量	try one's best		医院	hospital
赏花	look at some flowers		就医	see a doctor
接触	contact			

语法 Grammar

1. 主要 (mainly)

This adverb indicates the key aspect of something.

Examples:

感冒**主要**在人多的环境中传播。

The common cold is mainly transmitted in crowded environments.

春季的疾病**主要**有感冒和过敏。

Diseases commonly found in spring mainly include the common cold and allergy.

2. 尽量 (as far as possible/try one's best)

This adverbial modifier denotes the meaning of "as far as possible".

Examples:

我们要**尽量**早睡早起，才能有健康的身体。

For better health, we should try our best to go to sleep early and get up early.

请大家**尽量**发表意见。

Please voice your opinion as fully as possible.

3. 万一 (in case)

This construction indicates the supposition of something happening. It is often followed by a suggestion or solution.

Examples:

万一你迷路了，记得打电话给我。

In case you get lost, remember to call me.

万一失败了，你也不要伤心。

In case you fail, don't be sad.

词组及常用句式
Expressions & sentence structures

词组 Expressions

各种传染病	various infectious diseases
多发季节	the season when diseases frequently occur
提前做好预防措施	take precautions in advance
受疾病的侵扰	get infected
常见的春季传染病	common infectious diseases in spring
人多拥挤	crowded with people
经空气传播	airborne disease
肠胃不适	gastrointestinal discomfort
最常见	the most common
人们常说的	what people often say
干燥红肿	dry, red and swollen
树立防病意识	establish an awareness of disease prevention
接种相应的疫苗	get the appropriate vaccination
抵抗传染病	fight against infectious diseases
最佳手段	the best means
调整作息	adjust work and rest
空气不流通	poor ventilation
个人卫生	personal hygiene
养成良好的卫生习惯	develop good hygiene habits
饭前便后	before meals and after going to the washroom
风吹日晒	exposure to wind and sun
戴口罩	wear a mask
一旦感染流感	once infected with influenza

接受治疗	receive treatment
万一过敏发病	in case of an allergic reaction
刺激性食物	irritating foods
必要时	when necessary
到医院就医	see a doctor

常用句式 Sentence structures

1. 受 (sth.) 的侵扰

Explanation	Example
be affected by	很多老年人会受疾病的侵扰。 Many old people suffer from diseases.

2. 经 (sth.) 传播

Explanation	Example
spread through	病毒可以经空气传播。 Virus can spread through the air.

3. 尤其是

Explanation	Example
especially	天气一变冷，很多人会感冒，尤其是体弱者。 Once the cold weather comes, many people will catch a cold, especially those who are weak.

Option topic 4

Leisure

选修主题 4

休闲

4.1 娱乐消遣
Entertainment and pastimes

关键字 Key words

娱乐	消遣	周末	放假	方式
媒体	节目	音乐	影响	轻松

模拟试题 Mock questions

SL	1. 请你介绍平时的娱乐活动。
	2. 请说说你的假期生活。
	3. 请介绍一部你喜欢的电影。
	4. 你参加过夏令营吗？请介绍一下夏令营的活动。
	5. 采访一位明星，请他／她分享成功的经验。
HL	1. 请谈谈电视节目对青少年的影响。
	2. 请你邀请朋友参加某种娱乐俱乐部，并说说参加的益处。
	3. 请说说音乐对青少年的生活和思想产生的影响。
	4. 请介绍一次让你印象深刻的演出。
	5. 请说说什么样的电视节目对年轻人有益。

范文 Sample essay

请你介绍平时的娱乐活动。（SL）

我喜欢的娱乐活动

每当周末或放假的时候，我总喜欢坐在沙发上看各种电视节目或听不同的音乐。

电视节目有很多种，它们不但能带给我很多知识，而且能放松我的心情。比如纪录片和新闻等节目，我可以从中了解到以前和现在发生在世界各地的事情。而科幻片和卡通片能带我走进奇妙的世界，让我忘掉生活的烦恼。在众多电视节目中，我最喜欢的是喜剧片和功夫片。因为喜剧片能让我心情很愉快，功夫片则让我感觉很刺激。

除了看电视，我还很喜欢听音乐。和电视节目一样，音乐也有很多种，其中包括古典音乐和流行音乐。所有的音乐我都非常喜欢，我会根据不同的心情听不同的音乐。如果在学习中遇到了压力，我就会听轻松的音乐来缓解压力。如果心情不好，我会听节奏快的音乐，这样心情会好一点。音乐不仅带给我快乐，还带给我安慰，这就是我喜欢听音乐的原因。

虽然还有很多其它的娱乐方式可以选择，但是我最喜欢的仍然是看电视和听音乐。

Introduce two hobbies that you enjoy in your leisure time.

Talk about different kinds of TV programs and why you like them.

Talk about different kinds of music and when you will listen to them.

Summarize the passage.

(352 words)

词汇 Vocabulary

周末	weekend	从中	from	刺激	excited
放假	have a vacation	了解	understand	除了	except
电视	TV	以前	before	其中	among them
音乐	music	现在	now	包括	include
每当	whenever	发生	happen	古典	classical
总	always	纪录片	documentary	流行	popular
沙发	sofa	科幻片	science fiction	所有	all
各种	various	卡通片	animation	根据	according to
节目	program	功夫片	kung fu film	压力	pressure
听	listen to	奇妙	wonderful	缓解	ease
不同	different	世界	world	节奏	rhythm
不但	not only	忘掉	forget	安慰	comfort
带给	bring	生活	life	原因	reason
知识	knowledge	烦恼	trouble	虽然	although
而且	moreover	众多	many	其它	other
放松	relax	喜剧片	comedy	娱乐	entertainment
心情	mood	愉快	happy	方式	way
新闻	news	感觉	feel	仍然	still

语法 Grammar

1. 则 (yet/while)

"则" is used as an adverbial modifier. It is used in the second clause to indicate that the condition is not the same as what was mentioned earlier. It is placed after the second subject.

Examples:

喜剧片能让我心情很愉快，功夫片则让我感觉很刺激。

Comedies make me happy, while kung fu movies bring me excitement.

他是个非常诚实的人，她则不一样。

He is quite honest, yet she is not the same.

2. 除了……（以外），还／也…… (in addition to/also)

This construction indicates that what is mentioned in the second clause is also true, in addition to what has been mentioned in the first clause.

Examples:

他除了学习画画以外，还学习跳舞。

In addition to drawing, he is also learning to dance.

除了我和弟弟，爸爸妈妈也会去。

In addition to my brother and me, my parents will also go there.

3. 仍然 (still)

This adverb expresses the idea that a certain situation has lasted for a period of time or remains unchanged.

Examples:

昨天晚上十点的时候，他仍然在学习。

He was still studying at 10 o'clock last night.

虽然有很多种娱乐方式，但我仍然最喜欢钓鱼。

There are many kinds of entertainment, but I still like fishing the best.

词组及常用句式
Expressions & sentence structures

词组 Expressions

坐在沙发上	sit on the sofa
看各种电视节目	watch a variety of television programs
听不同的音乐	listen to different kinds of music
带给我很多知识	bring me a lot of knowledge
放松我的心情	relax my mind
从中了解到	learn from
发生在世界各地的事情	things happening around the world
带我走进奇妙的世界	take me into a wonderful world
忘掉生活的烦恼	forget the troubles of life
在众多电视节目中	among the many television programs
让我心情很愉快	make me feel very happy
让我感觉很刺激	make me feel very excited
和电视节目一样	same as television programs
根据不同的心情	depending on the mood
在学习中遇到了压力	encounter pressure while studying
听轻松的音乐来 缓解压力	listen to relaxing music to relieve stress
听节奏快的音乐	listen to fast tempo songs
带给我快乐	bring me joy
带给我安慰	bring me comfort
喜欢听音乐的原因	the reason I like listening to music
很多其它的娱乐方式	many other forms of entertainment

常用句式 Sentence structures

1. (sb.) 总

Explanation	Example
somebody always does something	我总喜欢坐在沙发上看各种电视节目或听不同的音乐。 I always like to sit on the sofa and watch TV or listen to all kinds of music.

2. 带给 (sb.) (sth.)

Explanation	Example
bring somebody something	音乐能带给我快乐。 Music can bring me joy.

3. 在 (sth.) 中

Explanation	Example
among something	在众多电视节目中，我最喜欢的是喜剧片和功夫片。 Among the many types of TV programs, I like comedies and kung fu films most.

范文 Sample essay

请谈谈电视节目对青少年的影响。（HL）

电视节目对青少年的消极影响

Give a general introduction of the negative influence of some TV programs on teenagers.

随着传媒的飞速发展，电视节目也以千奇百怪的姿态出现在观众的视野之中。这些节目有的乏善可陈，有的低级趣味，不仅对青少年没有任何教育意义，也给他们身心方面带来了消极的影响。

Talk about the first negative influence and give examples.

首先，冗长的电视剧分散了青少年的注意力，影响了他们的正常作息与学习生活。时下的电视剧，为了吸引观众、提高收视率，往往在每集的结尾处设置最有悬念的情节，令观众不由自主地一集一集追看下去。电视剧的精彩和轻松与学习的沉闷和紧张形成鲜明的对比，青少年不可避免地沉迷于电视剧中。这样不但影响了正常的学习生活，甚至还会影响身体健康。

Talk about the second negative influence and give examples.

其次，青少年心智未成熟，缺乏判断能力，不良的电视节目会向他们传递错误的资讯。现在不管是电视剧还是娱乐节目，内容大同小异。它们不是涉及商场的不良竞争，就是关于爱情的贪新忘旧。这些故事情节很多都是虚构的，但对于心智不成熟的青少年来说，却严重地影响了他们的价值观。除此之外，还有些电视节目涉及暴力、犯罪、吸毒等内容，这不但对青少年没有任何积极作用，反而诱导他们产生不良的行为。

Appeal to the whole society to be mindful of the negative influence of TV programs on teenagers.

电视节目的负面影响不容忽视，社会各界应加以重视，共同为青少年创造一个健康向上的生活环境。

(448 words)

词汇 Vocabulary

随着	with	影响	influence	鲜明	distinct
传媒	media	首先	first	对比	contrast
飞速	rapid	冗长	lengthy	不可避免	inevitable
发展	develop/ development	电视剧	TV series	沉迷	addiction
		分散	distract	甚至	even
千奇百怪	all sorts of strange things	注意力	attention	身体	body
		吸引	attract	健康	health
姿态	shape	提高	improve	其次	secondly
出现	appear	收视率	ratings	心智	mental
观众	audience	往往	often	成熟	mature
视野	vision	结尾	end	缺乏	lack of
乏善可陈	without anything good	设置	set	判断	judgment
		悬念	suspense	能力	ability
青少年	teenager	情节	plot	传递	transfer
任何	any	不由自主	involuntarily	错误	wrong
教育	education	精彩	wonderful	信息	information
意义	significance	轻松	relaxed	现在	now
身心	physically and mentally	沉闷	boring	不管	no matter
		紧张	nervous	内容	content
方面	aspect	形成	form	大同小异	more or less the same
消极	negative				

涉及	involve	犯罪	crime
商场	business sector	吸毒	take drugs
竞争	competition	积极	positive
关于	about	作用	role
爱情	love	反而	instead
贪新忘旧	indulge in the new one and forget the old one	诱导	induce
故事	story	产生	produce
虚构的	fictional	行为	behavior
对于	for	不容忽视	cannot be neglected
却	but	社会各界	different sectors of society
严重	serious	重视	emphasize
价值观	values	共同	common
涉及	involve	创造	create
暴力	violence		

语法 Grammar

1. 有的……有的 (some... while some)

In this structure, the pronoun "有的" functions as a modifier. It refers to part of a group of objects or people. The sentence is similar to "有些……有些……", but "有些" refers to more than one object or person, usually a large group. When both instances of "有的" modify the same noun, the noun is placed before the first "有的" and can be omitted in the second clause.

Examples:

这些节目有的乏善可陈，有的低级趣味。

Some of these programs have poor content, while others are simply vulgar.

有的书是买的，有的书是借的。

Some of the books are bought, while others are borrowed.

2. 不是……就是 (either... or)

This construction indicates that either statement may be true. The two phrases can be followed by nouns, verbs, phrases or clauses.

Examples:

他很喜欢音乐，每天不是听歌，就是弹琴。

He likes music very much. He is always either listening to music or playing the piano.

周末他们有很多计划，不是去公园野餐，就是去海边游泳。

They have a lot of plans for the weekend; they will either go for a picnic in the park or go swimming at the beach.

3. 不但……反而…… (on the contrary)

The conjunction "反而" serves as the turning point of the sentence. It is placed at the beginning of the second clause, while "不但" is placed in the negative clause containing either "不" or "没". The construction expresses the idea that a condition has caused a result counter to one's expectation.

Examples:

乏善可陈的电视节目不但对青少年没有任何积极作用，反而诱导他们做出不良的行为。

Bad TV programs won't bring about any positive influence on teenagers. On the contrary, they will induce teenagers to behave badly .

他不但不帮忙解决问题，反而故意给我们制造麻烦。

He didn't help to solve the problem. On the contrary, he deliberately caused trouble for us.

词组及常用句式
Expressions & sentence structures

词组 Expressions

随着传媒的飞速发展	with the rapid development of the media
以千奇百怪的姿态	in all sorts of strange manners
出现在观众的视野之中	appear to the audience
低级趣味	bad taste
带来了身心两方面的消极影响	negative impact to both physical and mental aspects
冗长的电视剧	long-winded TV dramas
分散了青少年的注意力	distract the attention of young people
影响了他们的正常作息	affect their daily routine
时下的电视剧	current TV dramas
为了吸引观众	in order to attract viewers
提高收视率	increase ratings
最有悬念的情节	a plot that creates the greatest suspense
一集一集追看下去	keep watching each episode
形成鲜明的对比	develop a sharp contrast
沉迷于电视剧	addicted to TV dramas
影响身体健康	affect one's
心智未成熟	immature
缺乏判断能力	lack of ability to judge
不良的电视节目	bad TV show
向他们传递错误的资讯	send them the wrong message
内容大同小异	more or less the same content
商场的不良竞争	harmful competition in business

影响了他们的价值观	influence their values
没有任何积极作用	no positive effect
不良行为倾向	bad behavioral tendencies
负面影响不容忽视	negative influences cannot be ignored
应加以重视	pay extra attention
健康向上的生活环境	healthy and positive living environment

常用句式 Sentence structures

1. 沉迷于 (sth.)

Explanation	Example
addicted to	很多年轻人沉迷于电视剧。 Many young people are addicted to watching TV dramas.

2. 向 (sb.) 传递 (adj.) 资讯

Explanation	Example
send a message to somebody	不好的电视剧会向人们传递错误的资讯。 A bad TV series will send a wrong message to people.

3. 不容忽视

Explanation	Example
cannot be neglected	电视剧的负面影响不容忽视。 The negative influence of TV series cannot be neglected.

4.2 体育运动 Sports activities

关键字 Key words

体育	运动	种类	健康	参加
身心	影响	方式	锻炼	放松

模拟试题 Mock questions

SL	1. 请你介绍不同种类的运动。
	2. 运动的好处是什么？请谈一谈。
	3. 运动对青少年有什么影响？
	4. 请你邀请同学参加运动会，并说明邀请的原因。
	5. 请介绍你喜欢的运动，并说明原因。
HL	1. 请谈谈运动的方式及好处。
	2. 谈谈体育锻炼的重要性。
	3. 请鼓励你的同学参加运动会，并介绍比赛的项目。
	4. 请介绍一次体育比赛的经历，并说说你的感受。
	5. 常见的运动有哪些？运动对人们的生活有什么影响？

范文 Sample essay

请你说说不同的运动种类。（SL）

运动的种类

只有经常运动，人的身体才能抵抗疾病，心理也才会更加健康。因此，我们应该多参加运动，保持健康的身心。 — Explain the importance of doing exercise.

运动有很多种，我们可以按照自己的喜好来选择。常见的运动有跑步、游泳、篮球、排球、足球、羽毛球、溜冰和爬山等等。其中，跑步是最方便的一种运动方式，不但可以自己一个人进行，还不用受地方的限制。另外，游泳是夏季最受欢迎的运动，既可以享受清凉，又可以锻炼身体。其它的球类运动也非常受欢迎，特别是受年轻人欢迎。因为他们不仅有很多时间去运动，也很容易找到同学一起打球，还可以用学校的体育馆，非常方便。到了冬天，我们还可以去滑雪和溜冰。当然，爬山是一年四季都很受欢迎的一种活动。因为爬山不但可以锻炼身体，还可以呼吸新鲜空气，使人放松心情。 — Describe different kinds of sports and their benefits.

运动对于每个人来说，都是非常重要的。因此，我们要经常运动，让自己拥有健康的身心。 — Summarize the passage.

(315 words)

173

词汇 Vocabulary

运动	sports	排球	volleyball	年轻人	young people
身心	physically and mentally	足球	football	因为	because
		羽毛球	badminton	容易	easy
健康	health	溜冰	skating	一起	together
只有	only	爬山	mountain climbing	学校	school
经常	often			提供	provide
抵抗	fight against	其中	among them	体育馆	gymnasium
疾病	disease	方便	convenient	非常	very
心理	psychological	一种	a kind	当然	of course
更加	more	方式	way	四季	four seasons
因此	so	不但 / 不仅	not only	锻炼	exercise/ physical training
应该	should	地方	place		
参加	take part in	限制	limit	身体	body
保持	keep	另外	in addition	呼吸	breathe
按照	according to	夏季	summer	新鲜	fresh
自己	oneself	受欢迎	popular	空气	air
喜好	preference	享受	enjoy	放松	relax
选择	choose	清凉	cool and refreshing	心情	mood
常见	common			对于	for
跑步	running	其它的	other	每个人	everyone
游泳	swimming	球类	ball	重要	important
篮球	basketball	特别	special		

语法 Grammar

1. 等等 (and so on/etc.)

This phrase is placed after two or more parallel items to indicate that the enumeration has not ended yet. There are more things that could be added after "等".

Examples:

我们去了很多地方，像北京、上海、杭州等等。

We went to a lot of places, like Beijing, Shanghai, Hangzhou, and so on.

爸爸买了很多家具，如椅子、桌子等等。

Dad bought a lot of furniture, such as chairs, tables, etc.

2. 特别是 (especially)

This adverb indicates that the following content stands out from what was mentioned before.

Examples:

球类运动非常受欢迎，特别是受年轻人欢迎。

Ball games are very popular, especially among young people.

中文对外国人来说很难，特别是写汉字。

Learning Chinese is very difficult for foreigners, especially writing characters.

3. 当然 (of course)

This adverb is used for emphasis or to indicate that what you are saying is obvious or generally known. It usually appears at the beginning of the sentence.

Examples:

当然，爬山是一年四季都很受欢迎的一种活动。

Of course, mountain climbing is a popular activity all year round.

我当然明白这句话的意思。

Of course, I understand the meaning of this sentence.

词组及常用句式
Expressions & sentence structures

词组 Expressions

抵抗疾病	fight against a disease
心理更加健康	better psychological health
多做运动	do more exercise
保持健康的身心	maintain healthy body and mind
按照自己的喜好来选择	choose according to your own preference
常见的运动	common sports
最方便的一种运动方式	the most convenient sport
自己一个人进行	do something by oneself
受地方的限制	limited by space
夏季最受欢迎的运动	the most popular summer sports
享受清凉	enjoy the cool breeze
锻炼身体	train your body
其它的球类运动	other ball games
非常受欢迎	very popular
特别是受年轻人欢迎	especially popular among young people
非常方便	very convenient
一年四季	in all four seasons
呼吸新鲜空气	breathe some fresh air
放松心情	relax your mind
对于每个人来说	for everyone
拥有健康的身心	have healthy body and mind

常用句式 Sentence structures

1. 按照 (sth.) 来选择

Explanation	Example
choose according to some criteria	我们可以按照自己的喜好来选择运动的方式。 We can choose different sports according to our own preferences.

2. 受 (sth.) 的限制

Explanation	Example
be restricted by something	在室内运动不用受天气的限制。 Indoor exercises are not restricted by weather.

3. (sth.) 非常受欢迎

Explanation	Example
something is very popular	游泳非常受欢迎。 Swimming is very popular.

范文 Sample essay

请谈谈运动的方式及好处。（HL）

运动的方式及好处

Introduce the importance of doing exercise.

运动与人们的生活息息相关，如果一个人每天都只是学习、工作，不做任何运动，那么，不仅他的身体健康会受影响，他的生活也会变得单调无味。

Encourage people to exercise.

只有经常参加体育运动，才能既锻炼身体，又丰富我们的生活。所以，我们不管再怎么忙，也要抽时间多做运动，否则，我们就不能享受美好的生活。

Talk about different kinds of sports and their benefits.

运动的方式有很多，虽然它们的进行方法很不一样，但都对我们有积极的影响。如果有很多人一起参与，又能找到运动场地，我们就可以打篮球、打排球或者踢足球。这些运动比较激烈，运动量比较大，能充分地锻炼身体。另外，和很多人一起运动，还能培养我们和他人的合作能力。而如果人数不多，就可以去打羽毛球、乒乓球或者网球。这些运动都非常有趣，也能很好地锻炼身体。当然，一个人也可以运动，我们可以选择跑步、跳绳或游泳。这些运动的好处是非常方便，只要有时间，就可以运动，不用受人数的限制。除了上面所说的各种运动，散步和爬山也是很受欢迎的运动。我们可以一边散心一边锻炼身体，有利于身心的健康。

Summarize the passage.

有这么多种不同的运动，我们的生活怎么还会单调呢？让我们一边锻炼身体，一边享受丰富美好的生活吧！

(419 words)

词汇 Vocabulary

方式	way/method		参与	participate in
生活	life		充分	fully
息息相关	closely related / an integral part of		另外	in addition
如果	if		培养	cultivate
那么	so		他人	other people
影响	influence		合作	cooperation
单调	dull		能力	ability
丰富	rich		有趣	interesting
不管	no matter		好处	benefit
抽时间	spare some time		人数	number of people
否则	otherwise		除了	except
美好	good		各种	various
虽然	although		爬山	mountain climbing
方法	method		有利于	beneficial to

语法 Grammar

1. 那么 (so)

This conjunction introduces the result or conclusion that is derived from the previous context.

Examples:

他不愿意别人帮忙。那么，让他自己解决吧。

He didn't want any help from other people. So, let him solve the problem by himself.

看来是没法出去了。那么，我们自己做饭吃吧。

It seems that we can't go out, so let's cook by ourselves.

2. 只有……才 (only... can)

This construction indicates that the subsequent result depends on the condition mentioned earlier."只有"can be placed before or after the subject in the first clause.

Examples:

只有经常锻炼，才能拥有健康的身体。

Only when we exercise frequently can we have healthy bodies.

我只有努力学习，才能考上理想的大学。

It is only by working hard that I can enter the ideal university.

3. 不管……都 / 也 (regardless)

This construction expresses the meaning that regardless of the situation, the result will not change. The point after"不管"should be in imperative form, expressing the notion "no matter what or how…". Either"都"or"也"can be used with"不管"in the sentence.

Examples:

不管再怎么忙，也不能忽略了运动的重要性。

No matter how busy we are, we should not undervalue the importance of playing sports.

不管明天天气怎样，我都会去见我的朋友。

Regardless of the weather, I will go out and meet my friend tomorrow.

词组及常用句式
Expressions & sentence structures

词组 Expressions

息息相关	closely related to/an integral part of
不做任何运动	do not do any exercise
身体健康	good health
受影响	be affected
变得单调无味	become very monotonous
经常参加体育运动	regular participation in sports
丰富我们的生活	enrich our lives
不管再怎么忙	no matter how busy
抽时间多做运动	find time to do more exercise
享受美好的生活	enjoy a beautiful life
运动的方式	type of sports
有积极的影响	have a positive impact on
很多人一起参与	with the participation of many people
运动场地	sports venue
比较激烈	comparatively more intense
充分地锻炼身体	fully train your body
培养我们和他人的合作能力	develop our ability to work with others
运动的好处	benefits of sports
有利于身心的健康	beneficial to physical and mental health

常用句式 Sentence structures

1. 与 (sth./sb.) 息息相关

Explanation	Example
an integral part of something	运动与人们的生活息息相关。 Sports activities are an integral part of our lives.

2. 变得 (adj.)

Explanation	Example
become	如果没有运动，生活将变得很单调。 Our lives would become very monotonous without sports.

3. 对 (sb./sth.) 有 (adj.) 影响

Explanation	Example
something has some effect on somebody or something	运动对我们有积极的影响。 Sports have a very positive effect on us.

4.3 旅行 Travel

关键字 Key words

旅行	旅游	景点	著名	吸引
游客	参观	了解	特色	安排

模拟试题 Mock questions

SL	1. 请你介绍一下中国著名的旅游景点。
	2. 请介绍中国的一个城市，吸引游客来旅游。
	3. 请说说你最难忘的一个地方，并解释为什么。
	4. 请说说你的一次旅游经历及感受。
	5. 旅游是一种常见的休闲方式，说说旅游有什么好处。
HL	1. 请你说说旅游前需要做哪些准备。
	2. 邀请你的朋友一起旅游，给他／她介绍你想去的景点。
	3. 暑假你想去一个著名城市旅游，说说你的旅游安排。
	4. 请介绍一次旅游的经历，并说说旅游的好处。
	5. 给你的朋友介绍中国的一个旅游城市。

范文 Sample essay

请你介绍一下中国的著名旅游景点。（SL）

中国著名的旅游景点

State that China has many famous scenic spots.

人们常常到中国旅游，因为中国有很多有名的景点，让人非常向往。

Introduce the most famous scenic spots in Beijing: the Great Wall and the Imperial Palace.

北京是很多人最常去的城市，最有名的景点有长城和故宫。长城有很多年历史，它很长，而且非常雄伟。而故宫是中国古代皇帝住的地方，有很多特别的建筑，非常有中国特色。所以，许多外国人很喜欢到北京旅游，参观这些著名的景点。

Introduce the most famous scenic spot in Xi'an: Terracotta Warriors and Horses.

另外，西安也是一个很有名的城市，它最著名的景点是兵马俑。兵马俑是中国第一个皇帝建造的，它们是用泥造的假人，大小和真人一样，排在一起像个军队一样，很有气势。很多游客都会到西安参观兵马俑，了解中国的历史。

Introduce the most famous scenic spot in Hangzhou: the West Lake.

杭州也吸引了很多游客，因为它的景色非常优美。春天是最适合去杭州旅游的季节，因为那时候不但天气很好，而且有很多花。杭州最有名的景点是西湖，人们如果去西湖游玩，不但能欣赏美丽的花，还能坐船游玩。

Invite readers to travel to China.

中国著名的旅游景点还有很多，如果想更了解中国，就赶快来中国旅游吧！

(336 words)

词汇 Vocabulary

常常	often		许多	a lot of
旅游	travel		外国人	foreigner
因为	because		参观	visit
有名	famous		著名	famous
景点	scenic spots		另外	in addition
向往	yearn for		兵马俑	Terracotta Warriors and Horses
北京	Beijing			
常去	often go to		建造	build
城市	city		军队	army
长城	The Great Wall		气势	impressive
故宫	The Imperial Palace		游客	tourist
历史	history		了解	understand
而且	moreover		吸引	attract
雄伟	majestic		优美	beautiful
古代	ancient		春天	spring
皇帝	emperor		适合	suitable for
特别	special		季节	season
建筑	building		天气	weather
特色	characteristics		欣赏	appreciate
所以	so			

语法 Grammar

1. 最 (most)

This adverb indicates the superlative degree among a group of things or people. It is usually placed before an adjective. It can also be placed before a verb denoting will or one's wishes, such as "喜欢"、"希望" etc.

Examples:

长城是北京最有名的景点。

The Great Wall is the most famous scenic spot in Beijing.

在这么多类型的歌曲中，我最喜欢摇滚乐。

Among different types of music, I like rock music the most.

2. 像……一样 (look like)

The construction indicates that something looks the same as something else. It is similar to "跟……一样".

Examples:

兵马俑排在一起像个军队一样。

The Terracotta Warriors and Horses line up together like an army.

那朵白云像个笑脸一样。

That white cloud looks like a smiling face.

3. 都 (all)

This adverb denotes the meaning of "all". It is usually placed after the subject.

Examples:

很多游客都会去西安参观兵马俑。

Many travelers will (all) go to Xi'an to see the Terracotta Warriors and Horses.

这些都不要紧，最重要的是要保证自己的安全。

All these matters are not important. The most important thing is your own safety.

词组及常用句式
Expressions & sentence structures

词组 Expressions

到中国旅行	travel to China
很多有名的景点	many well-known scenic spots
让人非常向往	make people yearn for
最有名的景点	the most famous attraction
有很多年历史	have a long history
非常雄伟	very majestic
中国古代皇帝住的地方	a place where the Chinese Emperor lived
特别的建筑	special architectural styles
非常有中国特色	full of Chinese characteristics
参观这些著名的景点	visit these famous attractions
排在一起	standing together in rows
像个军队	like an army
很有气势	very impressive
到西安参观兵马俑	visit the Terracotta Warriors and Horses in Xi'an
了解中国的历史	to understand Chinese history
吸引了很多游客	attract many visitors
景色非常优美	the scenery is very beautiful
最适合去杭州旅游的季节	most suitable season for traveling to Hangzhou
欣赏美丽的花	appreciate the beautiful flowers
中国著名的旅游景点	famous tourist attractions in China

187

常用句式 Sentence structures

1. 有……特色

Explanation	Example
have certain characteristics	这里的建筑非常有中国特色。 The architecture here is full of Chinese characteristics.

2. 到 (somewhere) 旅游

Explanation	Example
travel to somewhere	人们常常到中国旅游。 People often travel to China.

3. 去 (somewhere) 游玩

Explanation	Example
tour around somewhere	如果去了杭州，可以去西湖游玩。 If you go to Hangzhou, you can go on a tour around the West Lake.

范文 Sample essay

请你说说旅游前需要做哪些准备。（HL）

旅游的准备

旅游是一种很放松的休闲方式，能让人暂时忘掉学习生活的烦恼。但是，在旅游前也要做很多繁琐的准备，否则就不能享受旅游的过程。

Introduce the topic of making travel preparations.

首先，我们要确定旅游的地点。有些地方以看风景为主，有些以参观名胜古迹为主，有些以品尝美食为主，还有些以购物为主。我们要根据自己的喜好，来决定旅游的地点。只有选对了旅游的地点，才能达到旅游的期望。

List four different aspects of pre-travel preparation.

其次，我们要选择合适的交通方式。不同的交通工具能带来不同的旅游体验。如果乘坐火车，不但能欣赏途中的风景，还能享受舒适安全的旅程。但是，如果去国外旅行，乘坐飞机更加合适。虽然旅行的成本会增加，但是能节省时间。一旦确定了交通方式，最好提前订票，以免买不到票耽误了旅游计划。

还有，要合理安排好旅游路线和住宿。顺畅的行程才能让我们尽情享受旅游的乐趣。因此，我们在出发前要先做好行程安排，如计划好参观景点的顺序、旅游的时间分配和相应的住宿地点，确保自己的行程顺利而且安全。

当然，要是去国外旅游，还要提前办理签证手续。要确保自己的护照有效，并及时办理相关签证，避免签证的问题影响自己旅行。

Reiterate the importance of preparation before travel.

只要旅游前细心做好准备，就能拥有一个充实的旅程。

(429 words)

词汇 Vocabulary

放松	relax	合适	appropriate
休闲	leisure	交通方式	mode of transportation
方式	way	不同	different
暂时	temporarily	工具	tool
忘掉	forget	带来	bring
烦恼	trouble	体验	experience
旅游	tourism	乘坐	ride
繁琐	tedious and complicated	火车	train
准备	ready to	途中	on the way
否则	otherwise	享受	enjoy
享受	enjoy	舒适	comfortable
过程	process	安全	safe
首先	firstly	旅程	journey
确定	determine	国外	overseas
地点	site	旅行	travel
名胜古迹	historical sites	飞机	plane
品尝	taste	成本	cost
美食	food	增加	increase
购物	shopping	节省	save
根据	according to	一旦	once
决定	decide	最好	had better
达到	reach	提前	in advance
期望	expect	订票	reserve
其次	secondly	以免	lest
		耽误	delay

计划	plan
合理	reasonable
安排	arrange/arrangement
路线	route
住宿	accommodation
顺畅	smooth
尽情	enjoy
乐趣	fun
因此	so
出发	set out
顺序	order
分配	distribute/distribution
相应	corresponding
确保	guarantee
顺利	smooth

安全	security
当然	of course
要是	if
办理	deal with
签证	visa
手续	formalities
确保	guarantee
护照	passport
有效	valid
及时	in time
相关	related
问题	question
细心	careful
拥有	have
充实	full

语法 Grammar

1. 更（加）(more)

This adverb indicates a comparison between two things or people. It is usually placed before an adjective. It can also be placed before a verb denoting will or one's wishes, such as "喜欢", "希望", etc. The short form is "更".

Examples:

去国外旅行，坐飞机更加合适。

Taking a plane is more suitable when traveling overseas.

他比我更会说笑话。

He is a better joker than I am.

2. 以免 (in case)

This conjunction introduces the possible result under a particular condition.

Examples:

我们最好提前订票，以免票卖光了。

We had better book the tickets in advance, in case the tickets sell out.

我马上打电话给他，以免待会儿忘了。

I will call him right away, in case I forget later.

3. 要是 (if)

This conjunction is used in conditional clauses. It is often used with the adverb "就" to convey the meaning of "if... then...". The construction "要是……就……" functions the same as "如果……就……".

Examples:

要是去国外旅游，就要提前办签证。

If you travel abroad, you should apply for a visa in advance.

要是你不想去，就留在家里照顾弟弟吧。

If you don't want to go, then stay home and look after your brother.

词组及常用句式
Expressions & sentence structures

词组 Expressions

一种很放松的休闲方式	a very relaxing way
忘掉学习、生活的烦恼	forget the troubles of study and life
做很多繁琐的准备	do a lot of tedious and complicated preparation
享受旅游的过程	enjoy the process of traveling
确定旅游的地点	determine the travel destination
参观名胜	visit places of interest
品尝美食	enjoy delicious food
根据自己的喜好	according to your preferences
达到旅游的期望	achieve the expectations of the trip
选择合适的交通方式	select the appropriate mode of transport
不同的交通工具	different modes of transportation
带来不同的旅游体验	have a different travel experience
乘坐火车	take the train
欣赏途中的风景	enjoy the scenery along the way
享受舒适安全的旅程	enjoy a comfortable and safe journey
去国外旅行	travel abroad
乘坐飞机更加合适	taking a plane is more suitable
旅行的成本会增加	the cost of travel will increase
节省时间	save time
确定了交通方式	determine the mode of transport
最好提前订票	had better reserve tickets in advance
耽误了旅游计划	delay travel plans
合理安排	reasonable arrangement

顺畅的行程	smooth itinerary
尽情享受旅游的乐趣	enjoy the fun of travel
在出发前	before departure
先做好行程安排	plan the itinerary in advance
计划好参观景点的顺序	plan which scenic spots to visit first
旅游的时间分配	travel time distribution
相应的住宿地点	corresponding accommodation
确保行程顺利而且安全	ensure a smooth and safe trip
提前办理签证手续	apply for a visa in advance
影响了自己旅行	affect one's own travel
细心做好准备	carefully prepare

常用句式 Sentence structures

1. 以 (sth.) 为主

Explanation	Example
mainly for something	这个地方以看风景为主。 This place is mainly for sightseeing.

2. 达到 (sb./sth.) 的期望

Explanation	Example
meet somebody's expectations of something	我们做好旅游的准备，才能达到旅游的期望。 We have to prepare well for travel to meet our expectations for the trip.

3. 在 (sth.) 前

Explanation	Example
before doing something	在出发前，我们要做好计划。 We have to make good plans before traveling.

Option topic 5
Science and technology

选修主题 5
科学与技术

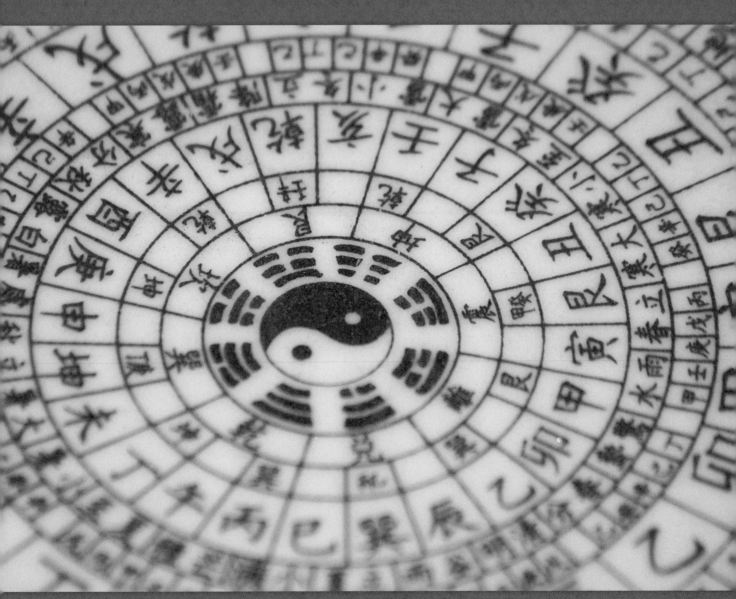

5.1 科技的影响
Influence of science and technology

关键字 Key words

科技	生活	影响	舒适	便利
丰富	负面	辐射	污染	危害

模拟试题 Mock questions

SL	1. 请你谈谈科技对生活的影响。
	2. 谈谈现代科技对青少年的影响。
	3. 现代科技为我们日常生活提供了哪些便利？
	4. 请谈谈电脑在青少年生活中的地位。
	5. 试谈谈科技发展和空气污染的关系。
HL	1. 你知道高科技会给环境带来哪些污染吗？
	2. 说说科技发展的负面影响。
	3. 谈谈高科技给人类生活的不良影响。
	4. 试分析科技对可再生能源的影响。
	5. 谈谈应该如何平衡科技发展及环境保护。

范文 Sample essay

请你谈谈科技对生活的影响。（SL）

科技对生活的影响

生活离不开科技。我们每天都要用的手机、电脑、网络，以及冰箱、电视等家用电器，这些都是科技的成果。科技对人类有非常大的影响。

科技让世界变小了。自从有了手机和网络，我们不管在哪里，都能和别人联系。另外，通过网络，我们能知道不同国家不同人的事情，不但增长了见识，而且拉近了和世界的距离。

科技让生活更加舒适便利。除了家里用的电器以外，便利的交通也提高了我们的生活水平。以前要花很长时间才能到达的地方，现在有了飞机和地铁，不但节省了时间，而且使旅程更加舒适。

科技让生活更丰富有趣。以前的娱乐活动很少，现在有了电视和电影，生活趣味大大提高了。而且随着科技的进步，越来越多的三维甚至四维电影出现，使我们的娱乐生活变得更加丰富有趣。

虽然科技带给我们很多方便，但人们在享受高科技产品的同时，也在承受着一些高科技产品所带来的负面影响，如辐射。辐射对于人的身体有害，这也是高科技在发展的过程中要面对的挑战。

(364 words)

Give a brief introduction of the impact of science and technology on our lives.

Talk about the first benefit of science and technology: they can draw people closer.

Talk about the second benefit: they can make life more convenient.

Talk about the third benefit: they can enhance our lifestyle.

Remind readers of the negative influence of science and technology.

词汇 Vocabulary

生活	life	国家	country	娱乐	entertainment
离不开	indispensable	不但	not only	趣味	interest
科技	science and technology	增长	grow	大大	greatly
每天	every day	见识	knowledge	随着	with
手机	mobile phone	而且	moreover	进步	advance
电脑	computer	拉近	close	越来越多	more and more
网络	network	距离	distance	甚至	even
以及	and	舒适	comfortable	出现	appear
冰箱	refrigerator	便利	convenient	虽然	although
电视	TV	除了	except	方便	convenient
家用电器	household appliances	交通	traffic	享受	enjoy
成果	achievement	提高	improve	高科技	high-tech
人类	human	水平	level	产品	product
影响	affect/influence	以前	before	同时	at the same time
世界	world	花	spend	承受	bear
自从	since	到达	arrive	负面	negative
不管	no matter	飞机	plane	辐射	radiation
别人	others	地铁	subway	有害	harmful
联系	contact	节省	save	发展	development
另外	in addition	旅程	journey	过程	process
通过	through	更加	more	面对	face
		丰富	a wealth of	挑战	challenge
		有趣	interesting		

语法 Grammar

1. 以及 (and)

This conjunction connects parallel nouns or different items.

Examples:

书桌上有笔、课本以及笔记本。

There are pens, textbooks and notebooks on the desk.

这个周末他看了两场电影以及一本书。

He watched two movies and read a book this weekend.

2. 自从 (ever since)

This phrase introduces the time when some changes took place. It can be placed before or after the subject.

Examples:

自从有了手机,我们的生活变得非常方便。

Ever since we got cell phones, our lives have become very convenient.

他自从学会了游泳,就天天和朋友去海滩。

Ever since he learnt how to swim, he has gone to the beach with his friends every day.

3. 在……中 (during/among)

This adverbial phrase conveys the meaning of "during a certain process" or "among a group".

Examples:

在发展的过程中,我们会面临很多挑战。

During development, we will face a lot of challenges.

在所有学生中,运动最出色的是小王。

Among all these students, Xiao Wang excels in sports.

词组及常用句式
Expressions & sentence structures

词组 Expressions

离不开科技	cannot live without science and technology
家用电器	household appliances
科技的成果	the achievement of science and technology
巨大的影响	a huge impact
自从有了手机和网络	ever since mobile phones and networks were invented
不管在哪里	no matter where
和别人联系	contact with others
通过网络	through the network
增长了见识	increase knowledge
拉近了和世界的距离	bring the world closer together
更加舒适便利	more comfortable and convenient
便利的交通	convenient transportation
生活水平	living standard
节省了时间	save time
使旅程更加舒适	make the journey more comfortable
娱乐活动	entertainment activities
大大提高	greatly improve
随着科技的进步	with the progress of science and technology
科技带给我们很多方便	technology brings us more convenience
享受高科技产品	enjoy high-tech products
承受负面的影响	have negative impact on
发展的过程中	the process of development

常用句式 Sentence structures

1. 离不开 (sth.)

Explanation	Example
can't live without something/closely related	生活离不开科技。 Our lives are closely related to science and technology.

2. 拉近了 (sb./sth.) 和 (sb./sth.) 的距离

Explanation	Example
bring somebody closer to somebody or something	网络拉近了我们和世界的距离。 The Internet brings us closer to the world.

3. 通过 (sth.)

Explanation	Example
through something	通过网络，我们能知道不同国家的事。 Through the Internet, we can learn about things that happen in other countries.

范文 Sample essay

你知道高科技会给环境带来哪些污染吗？（HL）

Explain that science and technology will cause pollution.	**高科技污染**
	科技的进步给我们的生活带来了巨大的变化，但不知大家有没有想过，科技也给我们带来了很多特殊的污染，比如高科技污染。
Categorize high-tech pollution into industrial pollution, nuclear pollution and electronic pollution.	高科技污染主要有工业污染、核污染、电子污染等等。
Explain what industrial pollution is and its impact.	工业污染主要是指工厂排放出的废水和废气。这些废弃物严重污染了环境，使得我们的天空不再蔚蓝，海水也不再清澈。
Explain what nuclear pollution is and its impact.	核污染和电子污染则是近几十年来产生的新型污染。核污染是一些核物质产生的放射性元素对人体和环境产生的破坏。这种污染的杀伤力极大，也很难根除。核污染主要来源于一些军事行动、核试验，甚至是核电站的爆炸。当年美国在广岛投放的原子弹以及近年海啸造成的核电站爆炸，都使得日本深受其害。受到辐射污染的人们，身体会变得很差，还容易患上白血病等癌症。
Explain what electronic pollution is and its impact.	电子污染主要是电子产品如电脑、手机、电池等给环境造成的污染。由于现在的科技高速发展，人们经常更换电子产品，但是对于淘汰产品的回收利用却做得不好。一些发达国家甚至把电子垃圾运往中国和印度，使得这两个国家成为电子垃圾的最大受害者。
Call on readers to find a solution to the potential harm.	科技造福人类，同时又带给了我们许多潜在而长久的危害。人类应该尽快寻找更好的方法来应对这一新的挑战。

(431 words)

词汇 Vocabulary

巨大	huge	产生	produce	癌症	cancer
变化	change	放射性	radioactive	由于	because
特殊	special	元素	element	高速	high-speed
污染	pollution	破坏	damage	更换	replace
主要	main	杀伤力	force of damage	淘汰	weed out
工业	industrial	根除	eradicate	回收利用	recycle and reuse
核	nuclear	来源于	come from	发达	developed
电子	electronic	军事	military	印度	India
等等	etc.	行动	action	受害者	victim
工厂	factory	试验	test	造福	bring benefit to
排放	discharge/emissions	甚至	even	同时	at the same time
废水	waste water	爆炸	explosion	潜在	potential
废气	waste gas	当年	that year	长久	long
严重	serious	原子弹	atomic bomb	危害	harm
环境	environment	以及	and	人类	human
天空	sky	海啸	tsunami	尽快	as soon as possible
不再	no longer	造成	cause	寻找	look for
蔚蓝	blue	日本	Japan	方法	method
海水	sea water	容易	easy	应对	deal with
清澈	clear	患上	suffer from	挑战	challenge
几十年	decades	白血病	leukemia		

203

语法 Grammar

1. 使得 (make)

"使得" indicates a causal relationship. The common structure is "使得 + result".

Examples:

这些废弃物严重污染了环境，使得我们的天空不再蔚蓝，海水也不再清澈。

The wastes greatly damage our environment, making our sky no longer blue and sea water no longer clear.

这部电影使得大家都明白了一个道理。

This movie makes everybody understand a truth.

2. 由于 (due to/because)

This conjunction denotes a reason and is placed before or after the subject in the first clause, followed by the result clause. It can be used with "所以", "因此" or "因而".

Examples:

由于水被严重污染，河里的鱼都死了。

Because the water is severely polluted, all the fish in the river are dead.

由于他昨天生病了，所以没来参加聚会。

Due to the fact that he was sick yesterday, he didn't join the gathering.

3. 同时 (meanwhile/at the same time)

This adverb introduces a different aspect of a particular situation. The content can either be the opposite of what was mentioned previously or can provide further information to a previous point.

Examples:

科技造福人类，同时又带给人们许多潜在危害。

Science and technology bring benefits to mankind. At the same time, they also bring a lot of potential danger to humanity.

人们纷纷捐款，同时，很多物资被运往灾区。

People donated in succession. At the same time, a lot of goods were dispatched to the disaster area.

词组及常用句式
Expressions & sentence structures

词组 Expressions

科技的进步	advances in technology
巨大的变化	tremendous changes
特殊的污染	special pollution
工业污染	industrial pollution
核污染	nuclear contamination
电子污染	electronic pollution
工厂排放出的废水和废气	emissions of waste water and gas from factories
严重污染了环境	cause serious pollution to the environment
天空不再蔚蓝	the sky is no longer blue
海水也不再清澈	sea water is also no longer clear
近几十年来	in recent decades
产生的新型污染	newly generated pollution
放射性元素	radioactive elements
杀伤力极大	highly lethal
很难根除	difficult to eradicate
主要来源于	mainly comes from
军事行动	military operations
核电站的爆炸	explosion of nuclear power plants
受到辐射的人们	people exposed to radiation
容易患上白血病等癌症	prone to cancer such as leukemia
科技高速发展	technology develops at a high rate
更换电子产品	replace electronic products

淘汰产品的回收利用	recycling things that are no longer used
电子垃圾的最大受害者	the victims of e-waste
科技造福人类	science and technology benefit mankind
潜在而长久的危害	potential long-term hazards
尽快寻找更好的方法	find better ways as soon as possible
应对这一新的挑战	respond to this new challenge

常用句式 Sentence structures

1. 不再 (adj.)

Explanation	Example
no longer	空气污染让天空不再蔚蓝。 Air pollution makes the sky no longer blue.

2. 给 (sth.) 造成的污染

Explanation	Example
cause pollution to something	很多电子废弃品给环境造成的污染很大。 Many electronic waste products cause great pollution to the environment.

3. 造福 (sb./sth.)

Explanation	Example
benefit somebody or something	科技会造福人类。 Science and technology will benefit mankind.

5.2 信息技术对社会产生的影响
Social influence of information technology

关键字 Key words

信息	技术	社会	影响	依赖
沟通	科技	产品	负面	交流

模拟试题 Mock questions

SL	1. 现代年轻人过分依赖互联网，以致失去与人沟通的能力。你赞成这种说法吗？
	2. 谈谈使用手机的利与弊。
	3. 你注意到年轻人过分依赖科技和互联网的现象吗？谈谈你的看法。
	4. 谈谈信息技术对人们生活的影响。
	5. 谈谈信息技术发展给社会及个人产生的影响。
HL	1. 请谈谈现今人们普遍使用手机的影响。
	2. 谈谈互联网对青少年的学习和生活产生的影响。
	3. 谈谈现今网络在人们生活中的地位。
	4. 试分析信息技术对社会产生的正面及负面影响。
	5. 请谈谈使用网络与人沟通的利与弊。

范文 Sample essay

现代年轻人过分依赖互联网，以致失去与人沟通的能力。你赞成这种说法吗。（SL）

Define your position on the influence of technology products.

Elaborate on how information technology affects people's communication ability.

Elaborate on the difference between online communication and face-to-face communication. Reiterate your position.

科技产品的不良影响

现在的青少年经常使用科技产品，很少与人沟通，导致他们失去了沟通能力。我赞成这个说法。

现代大部分的年轻人都拥有电脑和手机，他们每天花很多时间在这些科技产品上，导致很少有时间和家人或朋友沟通。他们放学后，不去参加课外活动，而是立即回家玩电脑游戏。这样，他们就只能生活在电脑的世界，不能跟真实世界的人沟通了。最近，经常有新闻说有些年轻人只躲在房间里用电脑，而不出门。后来，他们慢慢地失去了沟通的能力，害怕和别人说话。有些甚至因为害怕外面的世界，不敢去剪头发！更别说出去工作了！

虽然有人说科技拉近了人与人的距离，人们可以利用电脑和住得很远的朋友聊天，但是，用电脑谈话和面对面与人谈话很不一样。网上聊天主要是打字，但是在真实世界谈话时，除了语言，我们还要看别人的表情、语气和说话的环境，而这些都是网上聊天没有的。所以，我觉得如果年轻人花太多的时间玩电脑，就真的会失去与人沟通的能力。

(360 words)

词汇 Vocabulary

青少年	teenager	课外活动	extracurricular activities	拉近	close
经常	often	立即	immediately	距离	distance
使用	use	世界	world	利用	use
科技	science and technology	真实	real	聊天	chat
产品	product	最近	recently	谈话	talk
电脑	computer	新闻	news	面对面	face to face
沟通	communicate/ communication	躲	hide	网上	online
导致	lead to	房间	room	主要	main
失去	lose	后来	later	打字	type
能力	ability	慢慢地	slowly	但是	but
赞成	agree	害怕	fear	除了	except
说法	statement	甚至	even	语言	language
现代	modern	因为	because	别人	other people
大部分	most	不敢	dare not	表情	expression
拥有	have	剪头发	haircut	语气	tone
每	each	工作	work	环境	environment
参加	take part in	虽然	although	觉得	think
				如果	if

语法 Grammar

1. 立即 (right away)

This adverb indicates that a certain action takes place immediately.

Examples:

放学后，他立即回家玩电脑游戏。

He went home to play computer games right after school.

我立即叫了一辆的士去医院。

I went to the hospital by taxi right away.

2. 这样 (this way)

This construction introduces the result of the previous content.

Examples:

不懂的生词要查字典，这样才能学会它的用法。

You should look up new words in a dictionary. This way, you can learn how to use them.

我们什么都别告诉他，这样他就不会伤心了。

Don't tell him anything. This way, he will not be sad.

3. 慢慢地 (gradually/slowly)

When this adverb is used to mean "gradually" (i.e. over a period of time or in stages), it is placed in a separate clause. When it means "slowly" (i.e. referring to speed), it is placed before the verb.

Examples:

慢慢地，他们失去了沟通的能力。

Gradually, they lost the ability to communicate with others.

我慢慢地走了过去。

I slowly walked there.

词组及常用句式
Expressions & sentence structures

词组 Expressions

经常使用科技产品	frequent use of technological products
很少与人沟通	rarely communicate with others
失去了沟通能力	lose the ability to communicate
赞成这个说法	in favor of this argument
拥有电脑和手机	own computers and mobile phones
和家人或朋友沟通	to communicate with family or friends
参加课外活动	participate in extracurricular activities
立即回家	go home immediately
玩电脑游戏	play computer games
生活在电脑的世界	live in the computer world
跟真实世界的人沟通	communicate with people in the real world
躲在房间里	hide in the room
害怕和别人说话	afraid of talking to others
外面的世界	the world outside
出去工作	go out to work
拉近了人与人的距离	shorten the distance between people
网上聊天	online chat
别人的表情	someone's facial expression
说话的环境	speaking environment

常用句式 Sentence structures

1. 和 / 跟 / 与 (sb.) 沟通

Explanation	Example
communicate with somebody	如果一直上网，就不能和家人朋友面对面沟通。 If we surf the Internet all the time, we can't communicate with our families and friends face to face.

2. 失去……的能力

Explanation	Example
lose the ability of something/to do something	他们经常上网，慢慢地，他们失去了沟通的能力。 They spend so much time surfing the Internet that they gradually lose the ability to communicate.

3. 更别说 (sth.) 了

Explanation	Example
let alone something	有些人不敢跟人说话，更别说是出去工作了。 Some people don't dare talk to others, let alone go out to work.

范文 Sample essay

请谈谈现今人们普遍使用手机的影响。（HL）

使用手机的负面影响

无论是在街上还是地铁、公车上，我们都能看到行人和乘客拿着一部手机，各自埋头专注，仿佛外界的一切事物都与他们无关。手机的普及无疑给人们带来了很大的便利，但它也是一个冷酷无情的武器，切断了人与人面对面交流的热情。

手机的使用占据了人们大部分的时间，从而减少了人们面对面沟通的机会，间接导致感情的疏离。随着信息技术的迅速发展，手机的功能越来越多，几乎所有的事情都能通过手机完成。除了正常的通讯，还可以上网查资料、玩游戏等。很多人在家里或家庭聚会上宁愿玩手机，也不跟家人亲戚沟通交流。长期如此，人们面对面沟通也变成一件奢侈的事情，人与人之间的情感也很难维系。

更有甚者，手机的广泛应用让人变得只关注自己的事情，不注重与外界的接触和交流。如果一个人一直对着手机屏幕，他还能关注到除了他以外的其它事情吗？长此以往，他只会形成一种以自我为中心的思想。手机的普遍使用让人们只关注自己感兴趣的事情，沉浸在自己的世界里。

手机的确为人们带来了沟通交流的危机，因此，我们要清楚地知道手机普及可能带来的负面影响，并通过自我提醒和约束，尽量减少它的不良影响。

Describe the popularity of mobile phones and talk about their negative influence.

Talk about the first negative impact: reduced communication and distanced family relationships.

Elaborate the second negative impact: users become self-absorbed and shut themselves out of the outside world.

Remind readers to beware of the negative impact of mobile phones.

(436 words)

词汇 Vocabulary

无论	no matter	机会	opportunity	奢侈	luxury
街上	on the street	间接	indirect	维系	maintain
地铁	subway	感情	feeling	广泛	widely
公车	bus	疏离	distanced from	应用	apply/ application
行人	pedestrian	随着	as		
乘客	passenger	信息技术	information technology	关注	concern
各自	on their own			注重	pay attention to
专注	focus	迅速	quickly	接触	contact
仿佛	as if	发展	develop/ development	一直	always
外界	outside			屏幕	screen
一切	all	功能	function	形成	form
无关	have nothing to do with	几乎	almost	思想	thought
		所有的	all	感兴趣	interested in
普及	popularity	通过	through	沉浸	submerge
无疑	undoubtedly	完成	complete	的确	indeed
带来	bring	除了	except	危机	crisis
便利	convenient	正常	normal	清楚	clear
冷酷无情	ruthless	通讯	communication	负面	negative
武器	weapon	上网	surf the Internet	提醒	remind
切断	cut off	查资料	search	约束	constraints
热情	enthusiasm	玩游戏	play games	尽量	try to
占据	occupy	聚会	gathering	减少	reduce
从而	thus	宁愿	would rather	不良	bad
减少	reduce	亲戚	relatives		

语法 Grammar

1. 仿佛 (seem/as if)

This construction introduces a judgment based on the previous content, giving the impression of being something. The adverbs "似乎" and "好像" also have a similar meaning.

Examples:

所有人都看着自己的手机，**仿佛**外面的世界与他们无关。

Everybody looks at their own cell phone all the time as if what happens in the world outside is none of their business.

他说得绘声绘色，**仿佛**见过真人一样。

He gave such a vivid description that it seems he had seen the real person.

2. 宁愿 / 宁可……也不…… (would rather... than...)

This construction expresses the idea that a particular choice is preferable to another. The content following "宁愿" is the preferred choice, while that following "也不" is the rejected choice.

Examples:

有些人**宁愿**玩手机，**也不**跟家人沟通交流。

Some people would rather play on their cell phones than talk with their families.

钓鱼太无聊了！我**宁可**在家里看书，**也不**去钓鱼。

Fishing is too boring. I would rather stay at home and watch TV than go fishing.

3. 更有甚者 (what's more)

This conjunction indicates that the content following is more serious or severe than what has been mentioned before.

Examples:

更有甚者，人们变得只关心自己的事情。

What's more, people become self-absorbed and only care about themselves.

更有甚者，标准一直处在变化之中。

What's more, the standard is always changing.

词组及常用句式
Expressions & sentence structures

词组 Expressions

埋头专注	immerse and concentrate on
外界的一切事物	everything outside
与他们无关	nothing to do with them
手机的普及	the popularity of mobile phones
带来了很大的便利	bring a lot of convenience
冷酷无情的武器	ruthless weapon
人与人面对面交流	communicate face to face
占据了人们大部分的时间	occupy most of the time
间接导致感情的疏离	indirectly lead to a distanced relationship
随着信息技术的迅速发展	with the rapid development of information technology
手机的功能	functions of mobile phones
通过手机完成	completed by mobile phones
正常的通讯	normal communication
上网查资料	search for information on the internet
跟家人亲戚沟通交流	communicate with family and relatives
变成一件奢侈的事情	become a luxury
手机的广泛应用	wide use of mobile phones
只关注自己的事情	only concerned with their own business
注重与外界的接触和交流	focus on exchange with the outside world
以自我为中心的思想	self-absorption
沉浸在自己的世界里	submerged in their own world

为人们带来了沟通交流的危机	bring people the crisis of communication
通过自我提醒和约束	through self-reminders and self-control
尽量减少它的不良影响	minimize its adverse effects as much as possible

常用句式 Sentence structures

1. 与 (sb.) 无关

Explanation	Example
have nothing to do with somebody	很多年轻人走在路上一直看着手机，仿佛外面的世界与他们无关。 Many young people keep their eyes on their mobile phones all the time when they are on the road, as if the world outside has nothing to do with them.

2. 长此以往，……

Explanation	Example
if things go on like this	长此以往，人与人之间的沟通会越来越难。 If things go on like this, communication between people will get worse.

3. 沉浸在……的世界里

Explanation	Example
submerge in one's world	他经常沉浸在自己的世界里。 He is usually submerged in his own world.

Core topic 1
Communication and media
核心主题 1
交流与媒体

1.1 互联网
The Internet

Standard level

互联网的使用 The use of the Internet

对于年轻人来说，上网是一种时尚。	For young people, surfing the Internet is fashionable.
互联网现在非常普遍，很多人都享受着互联网带来的方便。	As the Internet is very common nowadays, people enjoy the convenience it brings.

互联网的好处 The advantages of the Internet

互联网让我们能随时沟通，也增长了我们的见识。	The Internet allows us to communicate with others freely anytime, and it can also increase our knowledge.
互联网方便我们的学习，我们可以在网上搜索到很多学习资源。	The Internet can facilitate our learning by allowing us to search for many resources online.
另外，我们也可以在网上结交来自世界各地的朋友，分享心情，交流经验。	Furthermore, we can make friends from all over the world on the Internet and share our experiences.
如果我们闷了，可以上网听听歌、看看电影或电视剧，非常方便，而且又不用花钱。	If we are bored we can listen to songs and watch movies on the Internet, which is very convenient and does not cost anything.

互联网的坏处 The disadvantages of the Internet

有些年轻人长时间上网，不和朋友交流，和朋友的感情变得越来越差，甚至忘记怎么跟人面对面交流。	Some young people spend so much time on the Internet that they neglect their friends and may even forget how to communicate with people face to face.
有些人一整天都在互联网上玩游戏，既浪费了时间，又影响了作息。	Some people play online games all day long, which is a waste of time and will affect their rest.
互联网的确能给我们带来方便，但是如果不小心，这些方便就可能变成负面影响。	The Internet undoubtedly brings us great convenience. However, if we are not careful, it can pose disadvantages to us.
有些网上的资讯可能是伪造的，还有些学习资料也可能是错误的。	The Internet can contain information that is fabricated, while some learning resources can also be wrong.
网上认识的人也有可能是骗子。	People you meet online may also be frauds.

Higher level

互联网的使用 The use of the Internet

随着科技的发展，互联网的应用也越来越广泛。	As technology develops, the use of the Internet has become more and more widespread.
很多人都在网上看电影和电视剧，而不是去电影院看电影，或者用电视机看电视剧。	Instead of going to the cinema or turning on the TV, many people now watch movies and TV dramas online.
人们在享受互联网提供种种便利的同时，也面临着越来越多的问题。	As people enjoy the benefits brought by the Internet, we also face more and more problems at the same time.
很多人只看到互联网方便有利的一面，却忽视了它可能带来的负面影响。	Many people only see the benefits of the Internet and undervalue its negative impact.
我们应该正确使用互联网，既尽情享受它提供的帮助，又尽量避免它可能导致的一些不良后果。	We should use the Internet wisely so that we can enjoy its benefits and at the same time avoid the negative consequences of Internet usage.
如果我们正确地使用互联网，互联网就能体现它最大的价值。	If we use the Internet correctly, then we can experience its full power.

互联网的好处 The advantages of the Internet

互联网不仅能让我们快捷地获得有用的资料，还能扩展我们的知识。	The Internet not only allows us to obtain information easily, but it can also expand our knowledge.
网上的学习资源很多，能够促进我们的学习。	The Internet contains an abundance of resources that can enhance our learning.
网络能够方便不同地区的人沟通交流，开阔我们的视野，还能够增进彼此的感情。	The Internet can facilitate communication between people in different areas, broaden our horizons and improve our relationships with one another.
在网上我们也能认识到更多优秀的人，扩大我们的交友圈子。	We can meet some great people on the Internet and enlarge our social circle by making more friends.

很多人在网上分享有用的资讯，参考别人的经验有助于我们自己的进步。	Many people share information on the Internet. We can improve ourselves by learning from their experiences.
互联网能够让我们轻松地享受网上娱乐资源，放松疲劳的身心。	The Internet enables us to relax and enjoy a range of entertainment online.
现在网上购物非常普遍，能节省很多时间和精力。	Online shopping, which has become very popular, can save a lot of energy and time.
除了可以文字聊天，还可以视频通话，就算距离再远也能联络感情。	Besides text, we can also communicate using web cameras, hence allowing us to maintain contact with other people regardless of the distance.

互联网的坏处 The disadvantages of the Internet

任何事情都有正反两面，如果使用不当，互联网也会带来负作用。	As with everything else, the Internet may bring negative consequences if not used correctly.
网络上的娱乐资源会占用我们很多时间，除了影响我们的健康，还影响我们的学习和工作。	Online entertainment has used up a lot of our time, which not only affects our health, but also our study and work.
有些不良网站涉及色情、暴力，会误导青少年，影响他们的身心健康。	Some websites, such as those with violent or pornographic content, are harmful to teenagers and will have a bad influence on them.
网络也少不了虚假的资讯，导致人们上当受骗，损失严重。	People who fall prey to false and misleading information on the Internet may suffer serious consequences.
长期对着电脑，会忽略和人沟通的重要性，让人越来越不懂得如何与他人相处，严重影响社交生活。	Using the computer for long periods of time will affect one's communication skills. People will not know how to communicate with other people face to face, which will severely affect their social lives.

1.2 广告
Advertisement

Standard level

广告的种类 Different kinds of ads

我们常见的广告有电视广告、报纸广告、杂志广告和网络广告。	The most common commercials we see are on TV, newspapers, magazines and the Internet.
不但商家会拍广告，政府也会拍公益广告，教育民众。	Not only do businesses advertise, but the government also advertises for charity or to educate the public.
随着网络越来越受欢迎，很多商家都在网络上做广告。	As the Internet becomes more and more popular, many businesses are turning to advertising online.

广告的作用 The function of ads

广告是传播资讯的一种方式，通过广告，商家能很好地传递资讯。	Advertisements are a good way of conveying information. Businesses can transmit their messages easily through commercials.
广告可以让更多的人知道有什么新产品。	Commercials can allow more people to know about new products.
广告既能让消费者很快认识和了解商品的功能和使用方法，也能让商家推销自己的商品。	Commercials allow people to learn how to use new products and better understand them, as well as allow businesses to sell their products.
广告能激发消费者的需求，吸引越来越多的人来购买。	Commercials can appeal to consumers' needs, attracting more and more people to purchase the goods.
广告一般会用一些很独特的方式来吸引人，使消费者非常想要那些商品。	Businesses usually use special tactics in their commercials to increase consumers' desire to buy their products.

广告能提高企业的名气，让越来越多的人关注企业的产品。	Businesses can increase their fame through commercials, making more people pay attention to their products.
广告的作用的确很大，不但传递了资讯，而且增加了消费需求，它使经济越来越活跃。	The effects of commercials are obvious. They not only convey messages, but also increase the desires of consumers, thus making the economy more and more dynamic.

广告的影响 The influence of ads

因为有些商家夸大了产品的作用，所以很多消费者发现买回来的产品比广告里的差。	Some merchants exaggerate the effects of their products. Many consumers find that the products they bought are not as good as they were advertised.
企业之间竞争激烈，有些企业会通过夸张的广告来吸引消费者。	Due to fierce competition, companies use exaggerated commercials to attract consumers.
很多商家利用明星来吸引年轻人，但是明星代言的产品不一定就是品质好的产品，如果仅仅因为崇拜明星就买，可能会买来没用的东西。	Many merchants use pop stars to attract teens, but their products may not necessarily be of good quality. If one buys those products because of the influence of those stars, one may have bought something useless.
因为广告太多，所以消费者不知道如何选择。	Because there are so many commercials, consumers simply do not know how to choose.

Higher level

广告的种类 Different kinds of ads

广告扮演着越来越重要的角色，很多产品资讯都要依靠广告才能传播。	Commercials are playing a more and more important role; many businesses have to rely on commercials to make their products known.
我们常见街道上和商场里张贴了很多广告，吸引消费者的关注。	We see many commercials on the streets and in shopping malls calling for our attention.
广告是广泛地向公众传递资讯的宣传手段，因此，很多商家利用广告宣传和推销产品。	Commercials are used to convey messages to the wider public; therefore, lots of merchants use commercials to promote their products.
现在网络广告越来越流行，因为很多人几乎每天都会上网，所以商家就利用人们这个习惯来做网络广告吸引消费人群。	As online advertising is becoming more popular and people are going online every day, merchants take the opportunity to attract consumers through online advertisements.
电视广告的优点是它不但能够传递产品资讯，而且能够加深人们的印象。	The advantage of TV advertising is that not only can it convey information about the product, it can also leave a lasting impression on the viewers.
报纸广告有很多就业资讯，能帮助人们找工作。	Classified ads contain lots of recruitment information, which can help people find jobs.
公益广告一般是关于环保或者关爱的，教育人们要保护环境和关爱身边的人。	Public service advertisements are usually about environmental protection or caring for others, thereby educating people to protect the environment or care about others.

广告的作用 The function of ads

广告是广泛地向公众传递资讯的宣传手段，通过不同形式的媒体，广告能满足不同人的需要。	Advertising is a way to spread a message to the public via different forms of media. Advertising can satisfy the needs of different people.
通过广告，商家成功地推销了商品，增加了销量，也带来了巨大的利益。	Through advertisements, merchants successfully promote their products, increase sales, and generate huge benefits.

广告能增加产品的知名度。	Advertisements can increase the fame of the products.
有些广告用明星来做代言人，增加了品牌的可信度。	Some advertisements use pop stars as spokespeople to increase the credibility of their products.

广告的影响 The influence of ads

广告的确为商家带来了很多利益，但正因如此，我们应该提防它有可能带来的负面影响。	Advertisements surely bring many benefits to merchants. Because of this, we should beware of their negative effects.
有些商家为了吸引更多消费者，制造假的优惠资讯，欺骗消费者。	In order to attract more consumers, some merchants make up fake discount information to deceive consumers.
盲目地相信广告，会给自己带来巨大的损失。	Blindly believing in advertisements can bring one huge losses.
商家利用人们的心理，用优惠、赠送等手段来吸引消费者，但是商品的品质却很差。	Some merchants manipulate consumers' mentality by offering discounts, free gifts and so on to attract them into buying low-quality products.
现在的广告形式和花样都非常多，作为消费者，我们应该善于分辨广告的真假，提防不良商家的不实广告，避免给自己带来不必要的损失。	There are many forms of advertisements. As consumers, we should be careful in distinguishing between what is true and false and beware of misleading advertisements created by unscrupulous merchants, in order to avoid unnecessary loss.
青少年要学会理性购物，不能盲目地追求物欲。	Teens have to learn to purchase sensibly and avoid blindly pursuing materialistic desires.

Core topic 2

Global issues

核心主题 2

全球性问题

2.1 全球化
Globalization

Standard level

全球化与文化 Globalization and culture	
文化全球化是一种趋势，它代表着全世界的人联系越来越紧密。	The cultural globalization is a trend which shows that the connections between people are becoming closer and closer.
全球的流行文化越来越一致。	The global pop culture is becoming more unified.
它不仅包括饮食文化，还包括娱乐流行文化。	It not only includes food culture, but also entertainment and pop culture.
美国的饮食文化渐渐地在全球各地流行起来，比如快餐店和咖啡连锁店。	The American food culture has become popular around the world, for example, fast food restaurants and coffee chain stores.
美国的娱乐节目在全世界也很流行。	American entertainment programs have also gained popularity worldwide.
全球化不是说全世界所有东西都变得一模一样。	Globalization does not mean all things become exactly the same.
现在中国饮食在西方国家越来越受欢迎，很多国家都有中国餐馆。	With a growing popularity of Chinese food in western countries, there are more and more Chinese restaurants opening in different countries.

全球化与政治 Globalization and politics

联合国负责管理全球事务。	The United Nations helps to manage global affairs.
随着全球化发展，很多问题已非单一国家能解决。	As a result of globalization, there are many problems that cannot be solved by a single country alone.

全球化与经济 Globalization and the economy

随着通讯技术和交通越来越发达，全球出现了越来越多的跨国企业。	As communication technology and transportation advance, there are more and more multinational enterprises.
经济全球化能让世界各地都享用同样的产品和服务。	Due to economic globalization, people around the world can enjoy the same products and services.
我们都知道这间美国公司的电子产品非常畅销，世界各地的人们都很喜欢用。	The products of this American electronics company are very popular. The whole world loves them.

Higher level

全球化与文化 Globalization and culture

受趋向单一的全球文化影响，各地的生活模式、价值观等越来越相似。	Influenced by a unifying global culture, lifestyles and values around the world are becoming more and more similar.
文化全球化和文化多元化是两个不同的概念，文化多元化指各民族的文化都得到尊重并得到平等对待。	Cultural globalization and cultural diversity are two different concepts. Cultural diversity means that all cultures should be respected and receive equal treatment.
虽然很多人觉得"文化全球化"就是"西化"，但是其实西方国家不一定是全球化的主导。	Although people may think that "cultural globalization" is equal to "westernization", western countries do not necessarily dominate globalization.
不论政治、经济或文化层面的全球化现象，都会引致不少国与国或文化与文化之间的冲突。	Whether political, economic or cultural, aspects of the globalization phenomenon will create conflicts between nations and cultures.

全球化与政治 Globalization and politics

多边组织是指成员来自多个国家的组织，一般指国际政府组织。	A multilateral organization is an organization whose membership is made up of different countries and it generally refers to international governmental organizations.
多边组织为成员国提供平台，合作解决跨越国界的问题。	Multilateral organizations provide a platform for members to solve the problems caused by cross-border cooperation.
现时最重要的多边组织是联合国，它负责管理全球事务。	The most important multilateral organization today is the United Nations, which helps to manage global affairs.
多边组织越来越多，影响力也越来越大。	As there are more and more multilateral organizations, their influence is becoming even greater.

透过国际合作，各国可以解决很多跨国问题，如气候变化、国际救援等，以维持国际政治、经济秩序。	Through international cooperation, countries can work towards many global issues, such as climate change and international aid, so as to maintain international political and economic order.

全球化与经济 Globalization and the economy

经济全球化会使全球范围内的贫富差距进一步扩大。	Economic globalization will further widen the gap between the rich and the poor all around the world.
全球化能带来许多好处和机会，也必将造成许多混乱和不和谐。	Globalization can bring many benefits and opportunities, but will also cause chaos and disharmony.
在经济全球化过程中，各国经济的相互依赖加强，这就是出现全球性金融危机的原因。	Economic interdependence between nations increases in times of economic globalization. That is one of the causes of global financial crises.
无论是在商品往来、贸易服务，还是观念交流等方面，全球化都对不同国家的经济、社会、文化构成很大的影响。	Whether in commodity trade, trade service or exchange of ideas, globalization poses a great challenge to the economies, societies and cultures of different countries.

全球化与本土化 Globalization and localization

实践全球化的同时不应忽视"本土化"因素，否则就会招致失败。	We should not ignore "localization" factors in pursuing globalization, or else it will end in failure.
一些快餐店为了符合当地口味，改变了不同地区的食物。	To suit the tastes of local people, some restaurants adapt their menus according to different regional tastes.
这间快餐店成功传播到世界上许多地区的原因，在于能与当地文化进行有效的协调，从而体现"全球本土化"的结果。	This fast food restaurant has successfully spread to many areas of the world by blending in with the local culture, which is a reflection of the results of "global localization".
这间美式快餐店曾经因为华人的饮食习惯，推出晚市饭类套餐和早餐，除了力求迎合香港人的饮食习惯，更贴近香港特有的"茶餐厅文化"。	Inspired by Chinese eating habits, this American fast food restaurant once introduced dinner and breakfast rice sets in Hong Kong. In addition to catering to Hong Kong people's eating habits, it also became closer to "Hong Kong restaurant culture".

2.2 国际组织与国际问题

International organizations and issues

Standard level

救援组织与非政府组织 Aid agencies and NGOs (non-governmental organizations)	
国际救援组织是一个全球化的组织。	International aid agencies are a global organization.
国际救援组织帮助人们解决很多紧急情况。	International aid agencies help people in emergency situations.
国际救援组织中往往有会不同语言的工作人员，方便救援行动的顺利展开。	International aid agencies have staff members who speak different languages to facilitate rescue operations.
国际救援组织在全球范围内有几百家成员公司。	International aid agencies have hundreds of members all around the world.
国际救援组织中的工作人员大多是专业人士。	Many people who work in international aid agencies are professionals.
最早的非政府组织是国际红十字会。	One of the oldest NGOs is the International Committee of the Red Cross (ICRC).
世界上最重要的非政府经济组织是世界贸易组织。	The most important NGO in the world is the World Trade Organization (WTO).
随着全球化的发展，国际救援组织和非政府组织会变得越来越重要。	Due to globalization, international aid agencies and NGOs are becoming more and more important.

和平、战争与贫穷 Peace, war and poverty

世界和平对全世界的人都非常重要。	World peace is valued by all people in the world.
战争会带来很多问题,严重影响人类的生活。	Wars will bring about a lot of problems and severely affect people's lives.
不同的国家为了自己的利益而发起战争。	Different countries wage war for their own benefit.
有的国家为了争夺石油等资源而发起战争。	Some countries wage wars in order to compete for resources such as oil.
战争还会产生饥荒与贫穷。	Wars also cause famine and poverty.
贫穷的国家需要国际救援和帮助。	Impoverished countries need help from international aid agencies.
在一些贫穷落后的地区,人们的生活条件很差。	In some indigent areas, people live in very poor conditions.

Higher level

救援组织与非政府组织 Aid agencies and NGOs (non-governmental organizations)	
国际救援组织在社会中扮演着重要的角色。	International aid agencies play an important role in society.
为了提高解决问题的能力，国际救援组织的工作人员都有着专业的背景并掌握不同的语言。	People with professional knowledge and strong language abilities work in international aid agencies to help solve problems.
一些国际组织为旅游人士、涉外工人和跨国公司提供协助。	Some international organizations provide assistance for travelers, immigrant workers and multinational enterprises.
国际红十字会是最早的非政府组织，多次获得诺贝尔和平奖。	The International Committee of the Red Cross, one of the oldest NGOs, has been awarded the Nobel Peace Prize several times.
最早的红十字会主要救助战争中的难民。	At the very beginning, the Red Cross mainly helped war refugees.
世界贸易组织制定贸易原则，解决国际贸易争端。	The World Trade Organization (WTO) sets up the principles of world trade and resolves international trade disputes.
世界贸易组织被称为"经济联合国"。	The WTO is known as "The United Nations of Economy."
国际救援组织与非政府组织在世界上发挥着越来越重要的作用。	International assistance groups and NGOs are playing a more and more important role in the world.

和平、战争与贫穷 Peace, war and poverty

全世界的人们都向往和平、爱好和平。	All people in the world are in favor of and look forward to peace.
热爱和平的人反对战争和一切暴力的行为。	People who love peace are against wars and all other violent behavior.
自从人类出现以来，战争就从没停止过。	Since the dawn of humanity, wars have never ceased.
战争给国家和人民带来巨大的损失和灾难。	War causes huge losses and disasters to countries and their people.
战争的结束并不意味着伤害的结束。	The end of a war does not signify the end of the damage.
战争结束后，还会出现一系列的遗留问题，比如：孤儿、疾病、经济萧条等。	The end of war leaves behind a trail of problems, including orphaned children, outbreak of diseases and economic depression.
现代战争采用高科技武器，杀伤力更加强大，比如核武器。	High-tech armaments, such as nuclear weapons used in modern warfare, can cause mass destruction.
核武器对人类身体的伤害巨大，所以国际社会禁止使用核武器。	International communities forbid the use of nuclear weapons due to their devastating damage to the human body.
贫穷是指人类生活极度困难，资源紧缺。	People in poverty lead an extremely difficult life and face a scarcity of resources.
造成贫穷的原因有很多，比如政治、历史、地理、文化、战争等。	Many factors contribute to poverty, such as politics, history, geography, culture and war.
越来越多的慈善组织和个人为贫穷地区做了很多贡献。	More and more charity organizations and individuals are helping and making a difference to those in poor regions.

2.3 毒品
Drugs

Standard level

毒品问题 Drug problems

年轻人的周围出现了越来越多的诱惑，他们也越来越容易受到毒品的伤害。	Facing more and more enticement, young people become increasingly vulnerable to drug damage.
很多人都是因为好奇或者心情低落，而去尝试毒品。	A lot of people are driven by curiosity or emotional distress to try drugs.
毒品很容易使人上瘾，而且对人有很大的负面影响。	Drugs are highly addictive and have numerous negative effects.
很多人受朋友的影响去吸毒，不但伤害了自己，而且伤害了他人。	A lot of people are pressured by their friends into taking drugs, which not only hurts them, but also others.

毒品的危害 Dangers of drugs

只要吸一次毒，我们就会对毒品上瘾，想要戒掉就很难了。	It only takes one try to become addicted to drugs. Giving up drugs will be very difficult.
吸毒会让人没法集中精神、记忆力减退。	Drugs will make people unable to concentrate and harm their memory.
毒品对人危害很大，长期吸毒的人身体都会很差，例如，他们总是觉得很累。	The dangers of drug use are enormous. Long-term drug abusers have poor health; for example, they may suffer chronic fatigue.
更严重的是，吸毒会让人产生幻觉，甚至引发对他人的危害。	What's even more serious is that drugs can cause hallucinations, making people harm others.

毒品的危害 Dangers of drugs

除此之外，吸毒还会让人走向犯罪，影响人的一生。	In addition, taking drugs can lead people to commit crimes and affect their entire lives.
很多吸毒者用完了家里所有的钱去吸毒，就会去偷去抢，来满足自己的毒瘾。	Many drug abusers spend all their savings to buy drugs, and turn to stealing or robbing to satisfy their addiction.

呼吁 Appeal

年轻人不要轻信坏人，要小心毒品的伤害。	Young people should not easily trust bad people and must be careful of the harm of drugs.
我们要拒绝毒品，不能让毒品伤害自己和他人。	We should say no to drugs, and prevent drugs from harming ourselves and others.

Higher level

毒品问题 Drug problems

现在，青少年滥用毒品的问题已经非常严重。	The problem of drug abuse among young people has now become very serious.
在现今的社会，我们周围存在着各种各样的毒品。吸毒不仅严重影响个人的身体健康，还给家庭、社会带来了很大负面影响。	In today's society, a variety of drugs exists around us. Drugs not only seriously affect an individual's physical health, but also bring negative effects on his/her family and society.
很多年轻人因为受到朋友的影响和其他人的唆使，而沾染了毒品，最终让自己后悔一生。	Many young people fall prey to drugs due to the influence or incitement of friends, which they will regret all through their life.
毒品问题应该受到社会各界的关注。	Society should be aware of and concerned about the problem of drugs.

毒品的危害 Dangers of drugs

毒品的危害非常大。	The dangers of drugs are tremendous.
毒品无论对个人、家庭还是社会都有不可磨灭的巨大影响。	Drugs not only have a huge impact on individuals, but also families and society.
毒品不但严重损害身体机能，还会导致吸毒者思想迷糊、记忆力衰退。	Drugs not only seriously damage physical body functions, but also lead to impaired thinking and memory loss.
一旦吸毒上瘾，就会让人丧失自制力，甚至导致犯罪的发生。	Once addicted, drug abusers may lose self-control, leading to crime.
毒品既会让人的身体产生依赖，又会使人的精神产生依赖。	Abusers can develop a physical as well as psychological dependence on drugs.
有些人为了支付高昂的毒品费用，不惜做违法的事情，如果被警察抓到，就可能面临坐牢的命运。	To pay the high cost of drugs, some people even do illegal things, and risk imprisonment if they are caught by the police.

毒品的危害 Dangers of drugs

我们经常看到一些家庭，因为父母吸毒，不仅使孩子失去了父母的保护，还给孩子的心理造成很大的创伤。	In some families, parents who are drug users fail to protect their children, bringing them psychological damage.
有些吸毒者，丧失了理智，甚至做出伤害家人的行为，给自己和家人都带来没法磨灭的心灵创伤。	Some drug addicts lose their minds and even hurt their family members, causing great pain to themselves and their loved ones.
吸毒除了使人无法正常地生活，还导致吸毒者失去自己珍爱的亲人朋友，甚至给社会带来潜在的危害。	Drugs make it difficult for people to lead a normal life, causing them to lose their friends and loved ones, and even bring harm to society.
有些缺钱的吸毒者，会去偷或去抢别人的钱财，导致他人的损失，严重影响了社会治安。还有些人因为毒瘾难忍，会做出伤人或杀人的异常举动，让无辜的人受到伤害。	Running out of money, drug abusers may steal or rob other people, which seriously affects the order of society. Under the influence of drugs, some may commit inexplicable acts, such as wounding or even attempting murder on innocent victims.

呼吁 Appeal

我们不仅要远离毒品，还要提醒那些吸毒的人赶快戒毒。	Not only should we stay away from drugs, but we must also persuade those who are using drugs to kick the habit.
我们不能因为一时好奇就沾染毒品，否则会后悔终生。	We should not try drugs out of curiosity, otherwise, we will regret it our whole lives.
为了拥有健康的身心和幸福的生活，我们都应该自觉远离毒品。	To have a healthy body and a happy life, we should stay clear from drugs.
青少年要有积极的人生观，不要因一时的心情低落而利用毒品麻醉自己。	Young people should have a positive outlook on life and should never use drugs as a means to block out their negative feelings.

2.4 环境保护
Environmental protection

Standard level

环境问题 Environmental problems

环境正一天一天恶化。	Our environment is getting worse and worse.
为了自己的方便，人们乱扔垃圾，乱排废气，导致严重的环境污染。	For their own convenience, people drop litter and drivers emit waste gas without control, leading to serious pollution.
在恶劣的环境中居住，我们的生活会受到很大的影响。	As we live in such a poor environment, our lives will be greatly influenced.
人们不知道资源是有限的，因此一点都不珍惜。	People don't know that resources are limited and therefore they don't treasure them.
马路上、河里有各种生活垃圾，影响了我们的市容，也破坏了环境卫生。	There is rubbish on the road and in the river, which influences the appearance of the city and damages our environment.
很多人用完水就直接倒掉，浪费了水资源。	A lot of people drain water directly after use, which is a waste of water resources.

环保措施 Measures to protect the environment

我们要节约用水，循环利用水资源，避免浪费。	We should save water, reuse water and avoid wastage.
去近距离的地方应该尽可能选择步行，去远的地方则可选择乘坐公车。	If we are going somewhere near, we should walk, and take the bus if the place is far.

环保措施 Measures to protect the environment

积极回收垃圾，这样既不会破坏环境，又可以循环利用资源。	By actively recycling garbage, we can reuse resources and do good for the environment.
为了让空气更洁净，我们应该多植树种花。	We can plant more trees and flowers to freshen the air.

呼吁 Appeal

环境是我们生存最基本的条件，如果我们破坏了环境，就等于破坏了我们生存的条件。	A clean environment is the most basic condition for human life. If we destroy the environment, it means we destroy our living conditions.
只有学会垃圾分类回收，养成习惯，我们的生活环境和生活品质才会更加美好。	If we can make a habit of separating and recycling waste, our living environment and quality of life will become better.
请从身边的小事做起，善待身边的动植物，爱护身边的环境。	Let's start with small things. Take good care of plants and animals, and care for the environment around us.

Higher level

环境问题 Environmental problems

随着人们的生活水平不断提高，我们对地球的破坏也越来越严重。	As our living standards increase, we bring more and more destruction to the earth.
为了追求经济效益，人们过度地使用资源，任意地排放废弃物，导致了很严重的资源浪费和环境污染。	In pursuit of economic efficiency, people use everything excessively and create waste with little control. These actions have led to a very serious waste of resources and pollution.
环境问题越来越严重，影响了我们生活的方方面面。	Environmental problems, which affect all aspects of our lives, are becoming more and more serious.
我们周围的空气受到严重的污染，污染主要来自汽车排出的废气。	The air around us is badly polluted by motor vehicle emissions.
由于缺乏环境保护意识，人们制造了很多垃圾，又排放了很多废气，导致了严重的水污染和空气污染。	Due to a lack of environmental awareness, people create a lot of garbage and exhaust emissions, leading to serious water and air pollution.
环境污染最直接的后果是使环境的品质下降，影响人类的生活质量、身体健康和生产活动。	The most direct consequence of environmental pollution is a decline in the quality of the environment, human life, health and production activities.

环保措施 Measures to protect the environment

我们最好随身携带环保袋，避免浪费塑胶袋。	We should bring our own shopping bags to avoid wasting plastic bags.
"低碳生活"与我们息息相关，涉及衣、食、住、行各个方面。	"Low-carbon living" is closely related to our daily life, involving aspects such as clothing, food, housing and transportation.
汽车废气是空气污染的一个重要原因，因此，我们应该尽量少乘车，减少废气的排放。	Motor vehicle exhaust is a major cause of air pollution, so we should try to reduce emissions by cutting down the frequency of trips.

我们要节约用水，循环利用水资源，避免浪费。	We should save and reuse water to avoid wastage.
政府应该多支持环保。	The government should do more to support environmental protection.
垃圾，只有混在一起的时候才是垃圾，一旦分类回收就能产生价值。	Garbage becomes useless when it is mixed up. Once the items are classified for recycling, they become valuable.
废电池混在垃圾中，不仅污染环境，而且废电池中的物质会影响人的健康。	Throwing used batteries in the garbage will not only pollute the environment, but the substances in used batteries may also affect human health.
我们要积极参与植树种花，一起创造更美好的环境。	We should plant more trees and flowers to create a better environment.

呼吁 Appeal

让我们携手创造一个美丽的城市！	Let's create a beautiful city together!
让我们自觉行动起来，过"低碳生活"，珍惜地球资源，保护我们的家园吧!	Let's take action! Live a low-carbon life, cherish the earth's resources and protect our homes.
在生活中节能减排，抑制全球暖化，我们承担着共同的责任。	We all bear the responsibility of reducing carbon emissions in daily life and to stop global warming.
环保需要每一个人的参与，为了我们的生活品质和身体健康，大家一起行动吧！过环保生活，减少环境污染，保护我们的地球家园！	Environmental protection needs everyone's participation. For the quality of our life and physical health, we need to take action: live a green life, reduce environmental pollution and protect our planet!

Core topic 3

Social relationships

核心主题 3

社会关系

3.1 关系
Relationships

Standard level

亲情和友谊 Family and friendship

中国人很看重家庭，家人对于中国人来说非常重要。	Family values are of great importance to Chinese people. They value their families very much.
父母不仅每天照顾我们的生活，还关心我们的学习情况，非常辛苦。	It is not easy for parents to care for our general well-being and our school performance at the same time.
家人的照顾让我们享受非常舒适的生活。	Our family members take care of us so that we can enjoy a comfortable life.
朋友对于每个人来说都非常重要。	Friends are very important for us all.
我们每个人都需要朋友，朋友既可以和我们分享喜悦，又可以为我们分担悲伤。	Everyone needs friends. We can share our happiness and sadness with one another.

维持关系 Maintaining relationships

当产生矛盾的时候，我们应该告诉自己其实问题不是很严重。	When there is a conflict, we should tell ourselves that the problem is not very serious.
年轻人要懂事，要学会感激和体谅父母。	Young people need to learn to appreciate and understand their parents.
我们要经常和朋友沟通，了解他们的想法。	We should communicate with our friends often to know more about what they're thinking.
如果和朋友分开了，也要和朋友多联系，维持友谊。	Even if we are separated from our friends, we should keep in touch to maintain the friendship.

人际关系 Interpersonal relationships

作为青少年，我们和家人呆在一起的时间是最多的，也最容易因为生活琐事和家人产生摩擦和矛盾。	As teenagers, we spend a lot of time with our families. It's easy to get into arguments with them over small issues.
父母和子女两代人的想法相差很大，因此，常常因为意见不合而导致争吵。	Parents and children have very different ways of thinking, which can often lead to arguments.
吵架对双方的伤害都很大，不仅没有解决矛盾，还影响了双方的关系。	Quarreling not only fails to resolve conflicts, but also affects the relationship of both sides.
年轻人不是很成熟，在和朋友相处的时候，有时候因为不愿意迁就对方而引起争吵。	Young people are not very mature. Sometimes they get into quarrels with their friends because they're not willing to think from the point of others.
朋友间的争吵很伤感情，甚至有可能让我们失去珍贵的友谊。	Quarrels can harm friendships and may even lead us to lose friends.

Higher level

亲情和友谊 Family and friendship

在外面不管遇到什么困难和挫折，我们都能从亲人身上找到很多安慰和鼓励。	Regardless of the difficulties and setbacks encountered outside, we can find comfort and encouragement from our families.
朋友是我们人生中不可缺少的部分，朋友不仅能和我们分享喜悦和悲伤，还能给予我们支持和鼓励。	Friends are an indispensable part of our lives. Friends not only share joy and sorrow with us, but also give us support and encouragement.
分离很多时候无法避免，我们也会因此想念一直陪伴我们的朋友。	When separation comes, as it inevitably does, we will miss our friends who used to accompany us.
亲情和友谊是我们人生中很重要的部分，能陪伴我们走过低谷。	Love and friendship is a very important part of our lives that can help us through difficulties.

维护关系 Maintaining relationships

在与人相处的过程中，很多时候都需要迁就和妥协。	Compromising is part of every relationship and getting along with others.
相互的尊重和理解，不仅能化解一些矛盾，而且增进彼此的感情。	Mutual respect and understanding will not only resolve conflicts but also improve relationships.
我们作为子女，应该理解父母，并自觉要求自己，尽量让父母放心。	As children, we should understand our parents and try our best to ease their worries.
只有家庭成员相互理解和包容，才能构成一个美满和睦的家。	A happy and harmonious home is built upon the understanding and empathy of each individual family member towards one another.
适当的沟通交流能消除误会，增进了解。所以，有时间应该多和朋友沟通交流。	Appropriate communication can eliminate misunderstandings and enhance mutual understanding. Therefore, we should have more communication with our friends.

人际关系 Interpersonal relationships

现在有越来越多的独生子女，因为家人太过宠爱，所以变得很任性。	There is a growing number of only children. As their families spoil them and focus all their attention on them, they become very headstrong.
其实很多矛盾都源于两代人之间的代沟，父母和孩子的想法不同，所以才有了分歧。	Many misunderstandings are derived from the gap between the two generations. Difference in ideas and way of thinking give rise to arguments between parents and children.
朋友间爱好不同、兴趣不同都会引发矛盾和争执，缺乏尊重和理解往往会导致我们失去友谊。	As friends have different hobbies and interests, this may lead to conflicts and quarrels. A lack of respect and understanding may even cause us to lose our friends.
生气时说的话对别人可能是一种伤害，甚至可能会毁掉友谊。	What we say when we are angry might be hurtful to others and even destroy friendships.

中国的家庭 The Chinese family

受传统家庭观念的影响，几代人一起生活在中国是很常见的事情。	Due to traditional family values, it is very common for different generations to live together in China.
随着社会生活水平的提高，家庭结构也改变了。	With the improvement of living quality, the family structure has changed.
多数年轻人希望结婚以后和老人分开生活，这也是一部分老年人的愿望。	Most young people prefer living apart from the elderly after getting married; this is also the wish of some of the older generation.
和老人分开住不能说明中国人不再重视家庭了，实际上，分开生活的年轻人还是尊敬、照顾老人的。	Although young people live separately from the elderly, they still value the family, and respect and care about the elderly.
节假日，子女们会带着礼品去看望老人，与老人一起过节、度假。	During festivals, children will bring gifts and visit the elderly to celebrate with them.

3.2 教育
Education

Standard level

教育方式 Teaching methods

教育对于孩子的成长有决定性作用。	Education plays a pivotal role in children's growth.
不同家长对教育有不同的看法,因此导致了他们对孩子有不同的要求。	As parents have varying views on education, they will have different demands on their children.
有些家长要求孩子考试拿高分,有些家长更注重孩子的其它能力,比如沟通能力和合作能力。	Some parents expect their children to get high exam scores, while others focus more on their children's abilities, such as communication skills and how they get along with others.
与美国教育相比,中国的教育比较重视学生的成绩。	Compared with education in the United States, Chinese education puts more emphasis on students' test results.
很多家长都对孩子的期望很高,不但要求自己的孩子上大学,而且希望他们考上名牌大学。	Many parents have high expectations of their children and hope they get into not just any university, but one that is prestigious.
当孩子成绩不理想的时候,他们不仅会责骂他们,还可能要求孩子上补习班。	When children get poor grades, they will be scolded and even be forced to take tutorial lessons.
虽然适当的压力能够督促孩子学习,帮助他们提高学习成绩,但是如果压力太大,就会造成不良的影响。	Although a suitable amount of pressure may motivate children to learn and help improve their academic performance, excessive pressure will cause adverse effects.
美国的教育重视让孩子自由发展,着重发挥孩子的想象力。	Education in the United States attaches great importance to allowing children to develop freely and nurturing their creativity.

| 学生不应该只会考试，应该有业余爱好或者其它的特长。 | Students should not only be exam-taking robots, but should also develop other hobbies and skills. |
| 家长不需要采用逼和惩罚的方式教育孩子，应该给孩子空间和鼓励。 | Parents should not use punishment or force when educating their children. They should give them freedom and encouragement. |

教育制度 Teaching system

严格和宽松这两种不同的教育理念哪个更好？这是很多人非常重视也经常讨论的话题。	Is a strict or lenient approach better when it comes to education? This is a topic that is frequently discussed.
中国有严格的考试制度，每个学生从小到大都要参加各种不同的考试。	China adopts a strict examination system in which students have to take a number of examinations from a young age onwards.
严格的考试制度让很多学生变得只会死读书，缺少思考和创新能力。	A strict examination system suppresses the thinking and creativity of many students, turning them into study robots.

Higher level

教育方式 Teaching methods

家长的鼓励和肯定对孩子能起很大的正面作用，除了能让孩子们更有信心，还能增加他们的学习主动性。	Encouragement from parents can have positive effects on children, which not only include increasing their self-confidence, but also their motivation to learn.
如果家长责骂或者惩罚孩子，孩子可能会因为承受太大的压力而厌恶学习。	Children who are scolded or punished by their parents may dislike studying as they are under too much pressure.
只要孩子懂得学习的重要性，就会自觉地认真学习。	Once children understand the importance of learning, they will be proactive in their studies.
很多家长对孩子的期望很高，他们经常担心如果孩子学习成绩不好，就没法和别人竞争。	Many parents have high expectations of their children. They often worry that if their children have poor academic performance, they will not be able to compete with others.
很多家长不仅要求自己的孩子上大学，而且希望他们考上名牌大学，这主要是因为上名牌大学对以后找工作有好处。	Many parents expect their children to enter not just any university, but one that is elite. They believe that attending a famous university would help them find a good job.
当学习成为一种负担时，试问哪个学生还会愿意去学习呢？	When learning becomes a burden, who would want to learn?
在选择大学专业上，家长多半希望孩子学理科或者读医学院、法律学院。	Regarding the choice of a major, most parents want their children to study science, medicine or law.
在家庭教育方面，中国家长也许应该学习美国家长，把选择的权利还给孩子，给孩子更多的自由。	With respect to family education, Chinese parents can perhaps learn from their US counterpart, who give children the right of choice and much more freedom.
学生应该在轻松、愉快的环境中学习，而不应该让他们感到有太大压力。	Students should learn in a relaxed and pleasant environment, and not be placed under too much pressure.

教育制度 Teaching system

教育制度随着时间的发展，经历了一系列的调整和改革。	Over the years, the education system has undergone a series of adjustments and reforms.
升学的压力很大，使许多学生变成了只会死读书的考试机器。	As students progress in their studies and the pressure increases, they become exam machines.
考试不应该是唯一的评估方式，学校应该考察学生各方面的能力。	Examinations should not be the only form of assessment; schools should take into consideration the all-round ability of students.
我们应该培养学生独立思考的能力，并帮助他们建立完整的价值观。	We should develop students' ability to think independently and help them to establish true values.
学生仍旧摆脱不了"分数决定一切"的命运。	Students still cannot escape the notion that "results determine everything".
很多教育者提倡施行双语教育。	Many educators support bilingual education.
双语教学能够为学生提供全面的语言学习环境。	Bilingual education can provide students with a comprehensive language learning environment.
懂得双语能提升学生的竞争力。	Being bilingual can enhance students' competitiveness.

Exercise and sample essays

练习及范文

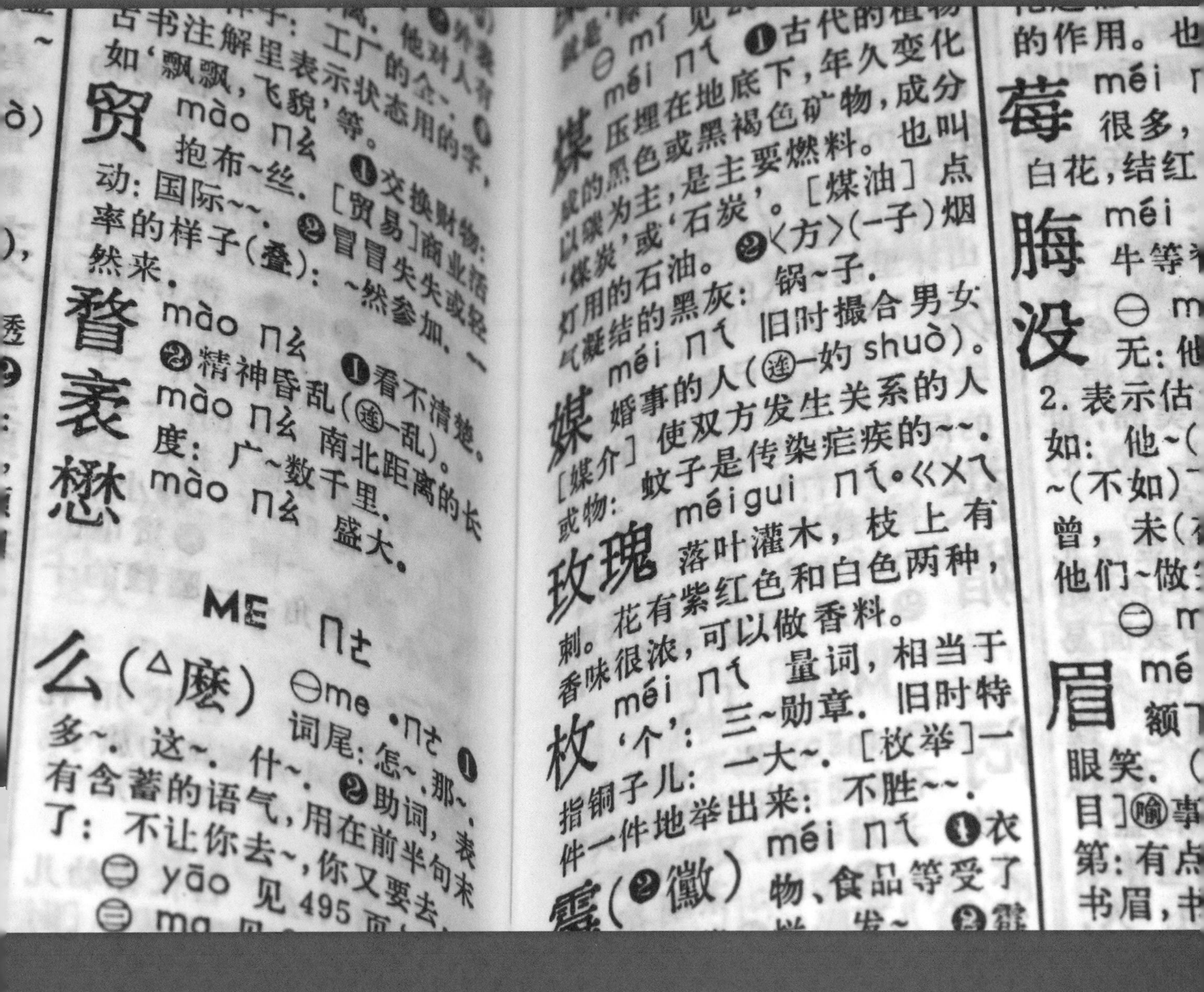

练习

SL试卷二

从下列题目中选择一题作答。字数在 300 字至480 字之间。

1. 文化多样性

1) 随着中国在国际上的地位逐渐提高，越来越多的人开始学习中文。请你写一篇博客，谈一谈学习中文的重要性。

2) 假如你是你们班中国文化学习小组的组长李山，将要负责举办一个"中国文化周"活动。你要作为代表在中文课发言，谈谈你们学习到的中国文化，并与西方文化进行比较，同时号召同学们积极参加文化周活动。请写这篇演讲稿。

2. 风俗与传统

1) 你的外国朋友梅西最近要去中国人家里作客，他担心自己不懂中国人的习俗于是向你请教。请写一封书信告诉他中国人的餐桌礼仪。

2) 为了加强学校各个国家学生的文化交流，最近学校校报组织了一次文化交流活动，请你给校报投一篇稿，介绍中国人的饮食习惯。

3. 健康

1) 今天你生病了没有去上学，请以日记的形式把你生病的起因和经历记录下来。

2) 中学生经常因为各种问题而导致身体不适，假如你是学生会会长孙海，你将要在学校的集会演讲，目的是提醒同学们要培养健康的生活方式。请写这篇演讲稿。

4. 休闲

1) 请给你的朋友写一封信，谈谈你的一次旅游经历和感受。

2) 你刚刚过了一个愉快又充实的暑假，请写一篇博客/部落格与网友分享你的假期生活。

5. 科学与技术

1) 请你以中学生的身份，在自己的博客上谈谈中学生使用手机的利与弊。

2) 学校的中文报征文，题目是《电脑在青少年生活中的地位》。写一篇文章投稿。

练习

HL试卷二A部分

从下列题目中选择一题作答。字数在 300 字至480 字之间。

1. 文化多样性

1) 你的朋友对中国菜很感兴趣，请给他写一封邮件，介绍一道有特色的中国菜式。

2) 现在越来越多的孩子从小就接受双语教育，请写一篇文章说一说双语学习有什么困难和优势。

2. 风俗与传统

1) 现在很多中国的年轻人喜欢西方的节日，而忽略中国的传统节日。针对这种现象，写一篇博客，说说你的看法。

2) 近段时间，学校出现了一些衣着不得体，过分攀比的学生。针对这种现象，学校即将进行一次主题为"学生该不该追求时尚"的演讲比赛，写一篇演讲稿，说一说你的看法。

3. 健康

1) 最近，你所在的城市爆发了一种流行病，假如你是学生会主席，请写一份注意事项告诉全校同学如何预防这种流行病。

2) 现在很多人的饮食习惯不健康。请用部落格/博文写一篇题为"饮食与健康"的文章，告诉大家健康饮食的重要性，帮助人们改正不良的饮食习惯。

4. 休闲

1) 请你给朋友写一封电子邮件，谈谈自己的一次旅游经历，并说说旅游有什么好处。

2) 你的表妹苏珊最近沉迷看电视，她的妈妈请你写一封电子邮件给苏珊，告诉她什么样的电视节目对年轻人有益。

5. 科学与技术

1) 请你写一篇博客谈谈使用网络与人沟通的利与弊。

2) 请你写一篇文章给校报投稿，谈谈科技发展的负面影响。

练习

HL试卷二B部分

1. "面对明星代言的广告商品，消费者要保持理性。"（广告）

2. "新闻媒体应秉承客观公正的原则。"（媒体的偏见）

3. "互联网对青少年的学习和生活影响非常大。"（互联网）

4. "手机已经成了人们生活中的必需品。"（电话）

5. "经常接触不健康的影视节目不利于青少年的健康成长。"（电视）

6. "远离毒品，珍爱生命。"（毒品）

7. "新能源的开发与利用已成为新世纪的研究焦点。"（能源储备）

8. "无论肤色、性别、宗教，所有的人都应该是平等的。"（种族主义、偏见、歧视）

9. "当今世界，全球化是不可避免的。"（全球化）

10. "人类改造大自然的同时，也对生存环境造成了严重破坏。"（人类对自然造成的影响）

11. "保护环境，人人有责。"（环境与可持续发展）

12. "解决代沟问题最重要的是尊重与理解。"（家庭关系）

13. "不同的国家拥有不同的文化和禁忌。"（社会禁忌）

14. "现在的教育正面临新挑战。"（教育制度）

15. "双语教育利大于弊。"（多语教育）

范文

根据下面的陈述，谈谈你个人的看法，并且做出解释和论证。你可以自由选择你所学过的任何一种文本类型作答。字数为180字至300字。

1. "网络购物的发展对人们影响很大。"

> 科技的发展一日千里，网上购物平台也如雨后春笋般地出现。网上购物的发展在各方面影响着人们的生活：它不仅增加了就业机会，还让人们的生活更方便多彩。
>
> 首先，网上购物增加了人们从商的机会。数不胜数的人通过网络平台找到了一个新的收入来源。行动不便的人或赋闲在家的主妇，足不出户便能开创一份事业，减轻家庭负担。
>
> 其次，网上购物的发展也让购物人士得到更多保障。以前人们担心网上购物会被窃取金钱，但现在有了担保交易和货到付款等方式，人们再也不用诚惶诚恐了。
>
> 再次，以低价为核心竞争力的网上购物，更让购物人士能以相对低廉的价钱购物。因此，收入不高的人也能在网上轻轻松松买个痛快。
>
> 总而言之，我们应珍惜网上购物发展所带来的机遇，充分发挥它的益处。

（287字）

2. "广告为商家带来利益的同时，也引发了一系列的问题。"

　　现今，各种广告铺天盖地，无处不在。大量的广告带来了丰富的资讯，也给商家带来了经济效益。然而，广告引起的一系列问题也不容忽视。

　　首先，有些商家为了追求自己的利益，夸大甚至捏造产品的作用。比如有些无良商家在广告中宣传自己的药品包治百病，而实际上却可能不仅无法治病，反而有害健康。还有些产品明明是本地生产，却宣称是澳洲或英国生产。这些虚假信息都严重损害了消费者的利益。

　　另外，不少明星为了赚代言费，不顾产品质量好坏。一些追星族为了支持偶像，无论明星代言什么商品，他们都会争相购买。如果这些产品出了问题，不仅商家需要负责，为其代言的明星也脱不了关系。

　　由此可见，广告是一把双刃剑，既创造了商业价值，也引发了社会问题。消费者应该理性应对。

（281字）

3. "现今青少年吸毒问题日益严重。"

　　青少年吸毒问题日益严重，甚至有贩毒集团把毒品引入学校，使不少学生染上毒瘾。我认为社会应更积极面对这个问题，防止更多青少年误入歧途。

　　首先，青少年吸毒问题已经到了刻不容缓的地步。一些学生被贩毒集团利用后，不但自己吸毒，还在校园贩毒。更有甚者成为贩毒集团的首脑，利用同学关系，把毒品卖入校园。因此，此问题已不容忽视。

　　另外，父母的疏忽也助长了青少年吸毒的问题，不少父母对孩子吸毒并不知情。如警方曾破获一个贩毒集团，抓了一名学生嫌疑犯。当学生被捕时，母亲竟对儿子吸毒、贩毒全然不知。

　　所以，面对青少年吸毒问题，全社会应携手解决。一方面，政府应更严厉监管毒品流通；另一方面，老师和家长要多关心学生；最后，学生应靠个人意志抵制毒品。只有这样，才能让我们的年青人有健康的未来！

（295字）

4. "环境问题与可持续发展是人类面临的重要挑战。"

当今世界，经济、科技、文化都在迅速发展，然而由此带来的污染却越来越严重。如何处理好环境问题，做到可持续发展，是我们人类必须面对的挑战。

环境污染威胁人类健康，降低我们的生活质量。众所周知，水和空气是人类赖以生存的重要因素。如果人类不能饮用干净的水，无法呼吸新鲜的空气，那么我们的健康就得不到保障。严重的空气污染还导致了温室效应，使得冰川融化，气候变暖，很多动植物濒临灭绝。

经济发展固然重要，环境问题却更不容忽视。一些工厂为了降低成本，将污水直接排入河流。这种只看眼前经济利益的做法不仅严重破坏了我们的生存环境，对我们的子孙后代更为不利。我们的地球已经岌岌可危。

为了保护我们的家园，坚持可持续发展是人类唯一的选择！

（276字）

5. "尊重父母应该是公民的重要责任。"

不管你身处哪个国家，不管你是什么身份，不管你有多大年纪，尊重父母都应该是我们公民的重要责任。

尊重父母是感激父母多年悉心栽培我们的一种方法。父母为我们流的每一滴汗水、每一滴泪水和付出的每一份心血，难道都是我们理所应得的？从婴儿呱呱落地开始，有谁没有感受过父母孜孜不倦的教诲？但社会中仍然有很多人不懂得珍惜父母的爱，不但窃取父母的财物，甚至还抛弃年迈的双亲。我们不应只顾自己的生活，而应懂得尊重、感激父母。

我们对父母的尊重不应仅限与父母一起庆祝节日。难道父母只在儿童节才爱惜照顾我们吗？我们应无时无刻都尊重他们。当父母年老时，更应无微不至地照顾他们，就如年少时他们爱我们一样。

总之，尊重父母是我们的重要责任。"百善孝为先"，让我们从现在开始行动——尊重父母，创造更美好的明天！

（303字）

6. "尊重与理解是良好人际关系的首要条件。"

　　处理好人际关系对于我们每一个人来说都至关重要。无论是面对家人、老师还是同学，甚至是陌生人，我们都应该做到尊重与理解。

　　在家庭中，尊重与理解是和睦关系的基础。有些孩子和父母的关系总是不融洽，常常产生各种矛盾。孩子认为家长不理解他们，家长也认为孩子不了解自己的良苦用心。如果父母与孩子能设身处地为对方着想，那么双方的误解就会少很多，家庭氛围也会更加温馨。

　　在学校中，尊重与理解有利于我们建立良好的师生关系和美好的友谊。当我们遇到矛盾时，应该认真聆听对方的想法，尊重对方的意见，并进行积极的沟通。老师与学生的关系也应该是平等互爱的，而不能把老师当作绝对的权威。

　　总之，如果想要拥有良好的人际关系，我们就要做到尊重与理解。

（279字）

Students' essays

学生作文

中文B SL

1. 每个人都有自己独特的爱好，有的喜欢运动，有的喜欢音乐。请你给你的朋友写一份电子邮件，与他／她分享你的新爱好，并谈一谈这个爱好有什么好处。

寄件人：lili@email.com
收件人：xiaoming@email.com
主题：我的新爱好

日期：2013年5月10日 星期五20:15

亲爱的小明： ┄┄┄┄┄

你好！最近怎么样？

我听说你最近**做羽毛球做的很努力**。我想跟你分享我的新爱好：排球。每星期三，我在学校跟同学们打排球。排球既有趣又好玩，它很吸引我。

搭配不当，应改为"打羽毛球打得很努力"。

打排球有很多好处：不仅**帮助你感压和感肥**，你还可以结交新朋友。排球是**一些团体运动，这些运动**会提高沟通和领导能力。

错别字，应改为"帮助你减压和减肥"。

"种"作为量词使用在"运动"前。应改为"一种团体运动，这些团体运动"。

有的学生很担心他们如果打排球或者做运动的话，就没有时间做作业。可是我觉得**做排球**既锻炼身体，又给学生更多体力读书，这也有利于学习。除了沟通和领导能力，排球还让你保持自信和减轻压力。这可以让学生的学习成绩进步很多。

打排球。

另外，**我们上的排球对是激烈的运动**，**所以**这种运动还可以帮我们减肥。打完排球以后还能放松心情。

句子表达不清晰，可改为"我们排球队经常训练"。

因此。

排球让我们的身体更健康，**和社交生活更丰富**。所以我打算长时间**做排球**。

打排球。

"和"也不可以直接用在分句的句首，可改为"也让我们的社交生活更丰富"。

你呢？羽毛球有什么好处？

打。

祝你生活愉快！

┄┄┄┄┄ *朋友。*

莉莉

评分标准 Benchmark*

评分标准 Benchmark	该项标准总分 Total sub-score	本文得分 Score in the student's essay
语言 Language	10	6
讯息 Content	10	7
形式 Format	5	4
总分 Total score	25	17

* 此评分标准只供参考，并不代表IBO官方立场。 *The scores are for reference only and do not in any way represent the views of IBO.*

教师点评

标准A语言：可以使用一些相关词汇，但有错别字及用词不当的现象，如："减"写成了"感"，"打排球"写成了"做排球"；简单句大致结构清晰，然而有些句子存在翻译痕迹，即英式中文，如："排球让我们的身体更健康，和社交生活更丰富"。

标准B讯息：内容明确，也较为丰富。文中提及打排球的不同方面的好处，但阐述不够具体。

标准C形式：基本符合电子邮件格式，但开头缺少称谓语，署名也不完整。

Teacher's comment

Criterion A – Language: Related words and phrases are used, but there are also some improper phrases and wrongly written characters, for example, "减" written as "感" and "打排球" written as "做排球". While the structure and meaning are clear in simple sentences, some sentences show signs of English influence, such as "排球让我们的身体更健康，和社交生活更丰富".

Criterion B – Content: The content is rich and clear. Some advantages of playing volleyball are mentioned, but there is a lack of details.

Criterion C – Format: Generally in accordance with the e-mail format, but the salutation is missing from the beginning and the signature is also incomplete.

2. 你的朋友写信说他／她很想在课余时间出去打工，可是他／她担心打工会影响学业，所以很想听听你的意见，请你写信告诉你的朋友你觉得应不应该在课余时间去打工，并告诉他／她原因。

"怎么"表示how，此处应使用"什么"。

语法不准确，可以改为"你只有十六岁，而且还在上学"。

特别疑问句的句尾应该使用"呢"。

改为"呢"。

改为"再"。

用词错误，应改为"进度"。

错别字，应为"期"。

用词不准确，可以改为"问题"。

用词不准确，可以改为"亲爱的思思"。

此处可以开始新的一段。

此封已经是回信。可以改为"希望于你有帮助。"

思思：

你好吗？很久没见！我听说你想找工作，你去找**怎么**工作**哪**？**你只是十六岁，你也有上学**，你怎么会有时间**哪**？工作有可能会影响你的读书**程度**。你可以在假**其**的时候打工或读完了中学**才**找工作吧！我觉得你现在最重要的事情就是努力地读书，得到一个好的分数，然后进入大学，选择一个你喜欢的科目。到时候，你去打工就没有问题啦！因为你进入了大学，你就不会有以前的**牵挂**啦！另外，你的父母又不是不给你钱，你有足够的零花钱。所以说，**老伴儿**，不要在读书的时候去打工！但是，如果你真的非常想打工，你就要小心选择打工的时间和打工的地方，还有打工的老板。最后我只想跟你说，我非常挂念你！我明年七月就会回到澳洲的！到时候见吧！

我的信先写到这里，**期待你的回信**。

祝你身体健康！

小敏上
四月十七日

评分标准 Benchmark*

评分标准 Benchmark	该项标准总分 Total sub-score	本文得分 Score in the student's essay
语言 Language	10	6
讯息 Content	10	6
形式 Format	5	4
总分 Total score	25	16

* 此评分标准只供参考，并不代表IBO官方立场。 *The scores are for reference only and do not in any way represent the views of IBO.*

教师点评

标准A语言：可以大致表达出写作意图，然而词汇较为简单，且存在用词不当的现象。如："老伴儿"实际是老年夫妻之间的互相称呼，用在朋友间是错误的。全文仅有290个字，未达到考试要求的300字，扣一分。

标准B讯息：一定程度上展现了自己的思想观点，可是文章内容不够充实，结构较乱，建议将正文内容分不同段落表达。

标准C形式：基本符合书信的写作要求，语气上还可以更加亲切。

Teacher's comment

Criterion A – Language: Overall fulfilled the purpose of writing, however, the vocabulary used is relatively simple and some phrases are used incorrectly. For instance, "老伴儿" is a term used between elderly couples, not between friends. With only 290 words, the text does not meet the 300-word requirement; 1 mark needs to be deducted.

Criterion B – Content: To some extent, the student can express his/her ideas, but the essay is poorly structured. Using separate paragraphs could enhance the organization. The content is also not quite substantial.

Criterion C – Format: Basically in accordance with the letter writing requirement; the tone can be warmer and more friendly.

3. 下星期是学校的科技周，这次科技周的主题是"科技与现代生活"。你要在科技周参加一个演讲比赛，题目是《互联网对青少年利大于弊》。请写这篇演讲稿。

尊敬的校长、各位老师、各位同学：

大家好！我是十一年级的方晓丽，今天我很雀跃，因为我能在这里和大家分享我对互联网对青少年的影响的感受。我演讲的题目是《互联网对青少年利大于弊》。

"雀跃"用在此处不合适，可以改为"荣幸"。

首先，互联网为青少年提供了新的渠道去取得各种资讯。现在有许多网站，例如"谷歌"、"雅虎"、"百度"等等。我们世界的科技非常厉害。学生只要上网一查就能找到足够的资料。我们不再使用传统的方法如去找资料。

如此表达太绝对，可以改为"我们可不再只使用传统的方法，如去图书馆找资料。"

第二，互联网有助于青少年不断提高技能。例如，青少年能够上新闻的网站，多了解新闻和意识到时事。这能提高学生的阅读技巧。

词组搭配不恰当，可改为"浏览新闻网站"。

"意识到"使用不正确，应该删去。

最后，互联网有助于开阔青少年的视野，加强青少年之间的交流和沟通。他们能够在聊天室、社交网站等等认识不同地方的文化、风土人情、传统和历史。

总括而言，互联网不但能为青少年提供新的方法去取得各种咨询，还可以令同学提高技能和开阔视野，所以我觉得互联网对青少年是利大于弊。谢谢大家。

可增加一句对同学们的呼吁："希望大家都能善用互联网！"

首行空两格。

评分标准 Benchmark*

评分标准 Benchmark	该项标准总分 Total sub-score	本文得分 Score in the student's essay
语言 Language	10	8
讯息 Content	10	8
形式 Format	5	4
总分 Total score	25	20

* 此评分标准只供参考，并不代表IBO官方立场。 *The scores are for reference only and do not in any way represent the views of IBO.*

教师点评

标准A语言：语句基本通顺，词汇的搭配上有些许轻微错误，如"我很雀跃"，"意识到时事"等。

标准B讯息：结构清晰，较好地传达了观点；内容上可适当增加互联网的坏处，使得文章更为全面。

标准C形式：基本符合演讲稿的写作要求和文体特点，如在文末加上一些呼吁或者鼓励性质的语句则更好。

Teacher's comment

Criterion A – Language: Sentences are generally fluent, but there are some wrong collocations, such as, "我很雀跃"，"意识到时事", etc.

Criterion B – Content: The structure is clear and views are conveyed. Some disadvantages of the Internet may be discussed to make the content more well-rounded.

Criterion C – Format: Basically in accordance with the writing requirement and stylistic features of speech notes. The closing may include an appeal to urge students to use the Internet more wisely.

4. 今天早上你的朋友告诉你一件事，让你很难过。请你写一篇日记，记录发生的事情，并说说自己的心里感受。

2013年6月15日　　星期六　　天气多云

今天早上，我认识了十年的朋友跟我说他有一件很重要的事要告诉我，原来他在月底就会离开香港，到美国读书。他说完后，我第一个反应是"嗯？"我记得他几个月前说他申请去哈佛大学，还跟我说他很希望自己能考上这一所大学。我很高兴地对他表达我为他的幸福，但同时我自己也有一点伤心，因为我知道我在香港会非常挂念他。

回到家，我躺在床上，很久都没动，一直在想我们认识对方的十年。我记得我第一次见到他的地方是在巴士上，当时他是坐在我的旁边。一开始，我们没怎么说话，但是因为我们差不多每天都见到对方，慢慢就成为了好朋友。做为他的朋友我很开心他能进哈佛大学读书，可是同时，我却不想他走。虽然我知道这种想法不合理，因为我不是永远见不到他，但是我无法摆脱这种感觉。

我最喜欢的一个回忆是我们一起去露营的一晚。一开始，我们有一点不开心，因为我们用了很长的时间去搭我们的帐篷，然后我们不能点火，因为天开始下雨，我们都有点失望。但是我们坐在帐篷里的时候突然见到月亮。我们安静地坐着，一直看着月亮。在那一刻，我觉得很宁静和舒服，因为只有我们两个坐在星空下。

想起这些回忆令我明白，其实无论他走或不走，他永远都是我的好朋友。因为无论他在哪里，我们一定会再次见面，而且我认识他那么久，一点分离没什么问题。

批注：

- 日记格式正确。
- 语法不正确，应改为："我很高兴向他表达了我的祝福"。
- 加入"当时"，则更好。
- "是"可删除。
- 错别字，应该为"作为"。
- 词语搭配不准确，可改为"我最难忘的一个回忆"。
- "突然"用在此处不恰当，可改为"竟然"。
- 句子表述不够准确，可改为"离别对我们来说不算什么问题"。

评分标准 Benchmark*

评分标准 Benchmark	该项标准总分 Total sub-score	本文得分 Score in the student's essay
语言 Language	10	8
讯息 Content	10	9
形式 Format	5	5
总分 Total score	25	22

* 此评分标准只供参考，并不代表IBO官方立场。 *The scores are for reference only and do not in any way represent the views of IBO.*

教师点评

标准A语言： 大都可以用准确的语法及词语来表达自己的内心世界，然而遣词造句方面还是有一些细微的错误。如"突然见到月亮"中"突然"的使用不恰当。

标准B讯息： 文章结构清晰，条理清楚。作者将自己即将与朋友离别的感情由浅入深地展现出来。

标准C形式： 无论是格式还是声调语气方面，作文都完全符合日记的要求。

Teacher's comment

Criterion A – Language: The student is largely able to use accurate grammar and vocabulary to express his/her feelings, although there are some minor mistakes in phrasing. For example, inappropriate use of "突然" in "突然见到月亮".

Criterion B – Content: The structure is clear and coherent. The writer expressed his/her emotions in parting with his/her friend in progressing levels from the surface to inner feelings.

Criterion C – Format: The student's writing style adheres to the style of diary both in the format and tone.

中文B HL A部分

1. 有个好朋友批评你的服装和发型不够时尚，你觉得很不开心。请你写一篇日记，说说发生了什么并谈谈你的心情。

二月三日　　　星期一　　　晴天

今天学校放假，我跟朋友去了看电影。我们看了近来热播的立体版《钢铁侠》，虽然我看了这部电影很多次，但是我还是百看不厌，这个爱情故事是我看过最具浪漫气氛的一部。

> 语法错误，应改为"我跟朋友去看了电影。"

看完电影后，我们去了吃晚饭，吃了我最爱的日本菜呢！让我吃得津津有味。正当我想站起来整理一下身上的衣服的时候，坐在对面的小欣做了一件令我非常愤怒的事——她说我穿的衣服和发型不够时尚！当时我非常低落，竟然一个朋友会这样跟我说话，让我在众人面前出丑。

> 语法错误，应改为"我们去吃晚饭"。

> 语序颠倒，应改为"一个朋友竟然会这样跟我说话"。

小欣批评我不要那么食古不化，还活在八十年代的日子，那时的我真的很想大发雷霆！她说我应该穿高根鞋，短群子，发型应该有点新意，才算时尚。她在众目睽睽之下批评我，令我满脸通红，哑口无言。身为我的好朋友，如果是为我好，不是不应该在众人面前批评我吗？

> 错别字，应改为"高跟鞋，短裙子"。

我跟小欣从小学到中学都那么好朋友，可以说是青梅竹马，她这样在众人面前批评我，我其实真的很愤怒。时尚这东西不能当饭吃的，更何况它日新月异，太难跟了。我觉得衣服最重要的是自己喜欢，自己穿得舒服，硬要跟着时尚而自己不喜欢那就没意思了。

> 语法错误，应改为"是很好的朋友"。

> 词语使用不当，改为"两小无猜"。

哎！我明白小欣这样批评我是出自好心的，但我觉得她应该想想我，在众人面前被自己的好朋友批评，那种感觉有多难受呢！好啦，我要收拾心情去睡觉了！明天再说吧！

评分标准 Benchmark*

评分标准 Benchmark	该项标准总分 Total sub-score	本文得分 Score in the student's essay
语言 Language	10	8
讯息 Content	10	6
形式 Format	5	5
总分 Total score	25	19

* 此评分标准只供参考，并不代表IBO官方立场。 *The scores are for reference only and do not in any way represent the views of IBO.*

教师点评

标准A语言：文中的四字词语是一大亮点，如"津津有味"、"大发雷霆"、"众目睽睽"等；然而，也存在有一些语法错误及错别字。

标准B讯息：文章内容松散，主题不突出。到底想表达"朋友间的相处问题"还是"对时尚的看法"呢？建议在构思文章时做到详略得当，突出中心。

标准C形式：格式及声调语气都符合日记的特点及要求，运用较多语气词，使得语言更加自然。

Teacher's comment

Criterion A – Language: Use of four-character idiomatic expressions, such as "津津有味", "大发雷霆" and "众目睽睽" makes the writing vivid. However, there are some grammatical mistakes and wrongly written characters.

Criterion B – Content: The content is loose and the main idea is not outstanding. It is not clear whether the student is expressing his/her views on "朋友间的相处问题" or "对时尚的看法". When brainstorming, think about the details that can be added that are relevant to the theme.

Criterion C – Format: The format and the tone are in accordance with the diary's requirement and stylistic features. Proper use of modal particles makes the language sound natural.

2. 你希望进大学后还能继续学习汉语，写一封信告诉你的笔友为什么你想继续学习汉语。

亲爱的小明：

　　你好吗？好久没通信了！听说你最近需要考试！希望你能**那**到好成绩！

　　其卖我有一件事想跟你分享：我打算在大学继续学习汉语。希望你可以给我一点你的想法。

　　第一，**我觉得我还没学好汉语，我只能说一点，听一点和写一点，我记得的词语也不够。**我希望好像一**人**真正中国人一样，**如果在中国生活我都不会有问题。**

　　第二，**在大学的汉语课上也会教一些中国历史。**虽然我是一个香港学生，但是我对中国的过去一点儿都不**认识。如果我想当做一个真正的中国人的语应该对中国历史有多点知识。**

　　其实，我希望在中国工作。**在加拿大，费用很高，我想多点儿学习中国文化。**

　　好了，我先说到这里吧！有空的话就回信吧！别太努力，记得要休息！

　　祝你

身体健康！

小善

三月二十日

左侧批注：

错别字，应为"拿"。

错别字，应为"其实"。

句子太长，显得累赘。应改为"我觉得我的汉语水平还不够高。我听说写汉语的能力有限，而且我平时积累的词汇也不足够"。

语法错误，缺少主语，应改为"我在大学的汉语课上还能了解一些中国历史。"

右侧批注：

量词搭配错误，应为"个"。

关联词使用不当，应改为"这样即使我在中国生活都不会有问题。"

用词不够准确，应为"熟悉"。

语句有毛病，应改为"如果想成为一个真正的中国人，我应该对中国历史有多点认识"。

意思表达不清，应加上一些关联词。可以改为"因为在加拿大的消费很高，所以我想去消费水平没那么高的中国工作。这样，我也应该提前学习多点中国文化知识"。

评分标准 Benchmark*

评分标准 Benchmark	该项标准总分 Total sub-score	本文得分 Score in the student's essay
语言 Language	10	5
讯息 Content	10	5
形式 Format	5	5
总分 Total score	25	15

*此评分标准只供参考，并不代表IBO官方立场。 *The scores are for reference only and do not in any way represent the views of IBO.*

教师点评

标准A语言： 学生语言能力一般，常出现错别字、用词不准确或语法错误的情况，比如"拿"写成"那"、"其实"写成"其卖"。同时，学生没有很好掌握句式的表达，比如复杂句不会用适当的关联词连接，简单句表意不清。另外，文章的词汇不够丰富，不会使用一些较难的四字词语，导致语言部分得分不高。全文仅有272个字，未达到考试要求的300字，扣一分。

标准B讯息： 文章提供的内容可以紧扣题目，而且会用一些连接词使得文章条理清晰，比如"第一"、"第二"。但是，文章思想观点的展开并不清楚，时有表意不清的情况发生。另外，对观点的支持性内容还不足够，使得文章传达的信息有限，比如可以多举例子来说明学习汉语的好处。

标准C形式： 整体来看，文章采用的文本类型清晰可辨。题目要求写一封信，文章的很多地方可以反映书信的文体特征，比如开头的称呼、结尾的祝福语、署名和日期等。因此，文章在形式部分的得分为满分5分。

Teacher's comment

Criterion A – Language: The student's language ability is mediocre; there are a number of wrongly written characters, inaccurate wording and grammatical errors, such as,"拿"written as"那"and"其实"written as"其卖". The student's sentences are weak: proper conjunctions are missing from complex sentences and the meaning is not clear even in simple sentences. There is a lack of variety in vocabulary, and four-character idiomatic expressions are not used, resulting in a low language score. With only 272 words, the text not meet the 300-word requirement; 1 mark will be deducted.

Criterion B – Content: The content is closely related to the subject, and some conjunctions are used to enhance clarity, such as"第一"and"第二". However, the ideas are not clear. In addition, the lack of supporting ideas limited the content. More examples can be given to illustrate the advantages of learning Chinese.

Criterion C – Format: Overall, the text type is clearly reflected. The question requires students to write a letter; many parts of the writing adhere to the stylistic features of a letter, such as salutation in the opening, blessing in the end, signature and date. Therefore, this part scored full marks.

3. 学校正举行一个辩论比赛，题目为《严格的老师能够帮助学生进步》。请你写一篇辩论稿说明你同意或是不同意这个观点的原因。

各位老师、各位同学：

此句多余，可删去。

观点太勉强，严格的老师不一定是用惩罚的方法来让学生学习的。可改为"各式各样有效的方法"。

"为此"用得好，可以再次重申论点。

老师对学生们的影响非常大。所以，一位老师的教学方式是非常重要的。在能不能够帮助学生的前提下，我认为严格的老师才能帮助学生进步。

首先，严格的老师才有能力维持学生们的专注力。学生们很多时候都会不小心在课堂上睡着。一个严格的老师能够用惩罚让他的学生们集中精神，甚至使他们不敢睡着。如果老师不严格，容许学生整节课都睡觉，那么学生将没有机会学习知识。

其次，一个严格的老师能使他的学生服从他。这样，学生们会由于害怕被老师骂而准时地完成他们的作业。完成作业绝对能够帮助学生们进步，因为知识需要"温故而知新"，只有通过做作业来复习，才能较好地掌握所学的知识。严格的老师能让自己的学生乖乖听话，而且准时交功课。

再次，一个严格的老师能设法让他的学生认真学习，专心地不和其他同学聊天地上课、工作。这样，学生们的学业必定能够进步。

最后，虽然一个严格的老师可能有时不能够让学生们快乐地学习，但是为了学生们的大好前途，这也是在所难免的。为此，我认为一个严格的老师能大大地帮助学生们的学业。

搭配错误，不说"维持学生们的专注力"，应改为"让学生们集中注意力，认真听讲"。

用词不准确，应改为"在课上打瞌睡"。

用词不够准确，可改为"听从他的指导"。

此处论点与第二段的观点重复，使文章内容重复累赘。应提出另一个观点

表意不清，应改为"老师的严格也是在所难免的"。

评分标准 Benchmark*

评分标准 Benchmark	该项标准总分 Total sub-score	本文得分 Score in the student's essay
语言 Language	10	7
讯息 Content	10	6
形式 Format	5	3
总分 Total score	25	16

* 此评分标准只供参考，并不代表IBO官方立场。 *The scores are for reference only and do not in any way represent the views of IBO.*

教师点评

标准A语言： 学生准确使用了一些范围广泛的词汇，但是还存在词语使用不准确和搭配错误的情况，比如"睡着"、"维持学生们的专注力"等。此外，学生在文章中有效地使用了一些复杂句结构，比如"虽然……但是"、"只有……才能"等。

标准B讯息： 文章的论点清晰，而且会使用一些关联词来使得思想观点的表达条理清晰，这一点做的比较好。但是，论证材料有重复出现的情况，比如文章第四段的分论点和第二段的分论点重复，导致不能很好地传达讯息。

标准C形式： 整体来看，文章采用的文本类型有时可以识别，但是写作惯用手法运用得比较有限。题目要求写一篇辩论稿，虽然文章正文里面能够看出作者的一个固定立场，但是缺少辩论稿的开头，比如"各位老师、各位同学"等。因此，文章在形式部分的得分为3分。

Teacher's comment

Criterion A – Language: The student shows mastery of a broad range of vocabulary, but there are some inaccurate words and mismatched phrases, such as "睡着，维持学生们的专注力", etc. Some complex sentence structures are used, for example, "虽然……但是"，"只有……才能", etc.

Criterion B – Content: The argumentation is clear and the use of connectives enhances clarity. However, there is repeated usage of supporting arguments, for example in the fourth and second paragraph, which weakens the content.

Criterion C – Format: Overall, the text type is reflected, but there is limited use of common writing tactics. The question requires students to write a debate script. The writer has a clear standpoint, but the text lacks a proper debate script opening, such as "各位老师、同学", etc. Therefore, this part scored 3 points.

4. 你是学校运动社团的社长，设计一份传单介绍你们运动社的活动，吸引新生加入你们的社团。

仁安学校运动社

各位新生：

俗语说："生命在于运动。" 运动是我们生活不可或缺的元素。它带给我们数不胜数的好处，有利于我们的身心健康，有助于纾缓我们在学习上的压力，而且能让我们看的年轻。所以，借着这次机会，本校运动社鼓励各位新生参加本社的活动，一起享受具有活力的人生吧！

活动日期：2014年2月10日至2014年5月30日

活动内容：

- 逢星期三放学后，有校车接送我们到湾仔运动场进行一千二百米径步。接着会举行一百和二百米赛跑。获胜者将获得神秘的奖品。时段为下午四时至六时，合共两小时。

- 逢星期五放学后，我们将留校，到学校游泳池游泳。游泳能够促进我们全身肌肉和四肢的发展，是一项非常全面的运动。时段为下午三时至下午四时半。

- 每隔两个星期，我们会举行一个"骑单车日"。我们会在星期六早上八时正在大围火车站E出口集合，步行至单车径，租用单车然后从大围骑单车到沙田为终点并交还单车。全程大约十六公里，需时约两小时。最后我们会到附近的餐厅享用丰富的午餐。

费用：费用全免（但不包括星期六的交通费）

名额：上限五十人（最少二十人）

联络方式 / 参加方法：可以向本校校务处索取表格，填好相关资料后交给校务处的工作人员即可。如有任何疑问请致电6532 1489（社长），或电邮至 sports@school.com。

开篇俗语的引用，可以为文章增色不少。

词语不够准确，应改为"显得"。

活动内容以点列式安排，使文章内容清晰，符合传单简洁有条理的文体特点。

错别字，应为"竞步"。

指出具体的"联络方式"，是传单写作的惯用手法。

成语的使用，也能让文章语言部分得分提高。

点出传单目的，很好地回应了题目"吸引新生加入社团"的目的。

搭配不正确，应为"举行一个'骑单车日'活动"。

运动有这么多的好处，赶快来参加吧！让我们享受健康和充满活力的人生！

此处缺少句子的主语，应改为"大家赶快来参加吧"。

仁安学校运动社社长

陈大明

2014年1月13日

评分标准 Benchmark*

评分标准 Benchmark	该项标准总分 Total sub-score	本文得分 Score in the student's essay
语言 Language	10	8
讯息 Content	10	9
形式 Format	5	5
总分 Total score	25	22

* 此评分标准只供参考，并不代表IBO官方立场。 *The scores are for reference only and do not in any way represent the views of IBO.*

教师点评

标准A语言： 学生比较有效地掌握了所学语言，可以使用各种复杂句式。虽然学生偶尔出现用词不准确的情况，比如"看的"应为"显得"，但是整体看来，学生的词汇比较丰富，还会使用恰当的四字成语和一些俗语，可见学生积累了较好的语言功底。

标准B讯息： 文章很好且有条理地传达了合适的思想观点，比如开头直接点明传单的目的是"鼓励学生参加活动"，紧接着以点列式清晰列出活动的细节和参加方式，结尾再次呼吁学生参加活动。整体看来，文章的支持性内容足够且合适，向读者有效传达了信息，达到了写作的目的。

标准C形式： 整体来看，文章采用的文本类型清晰可辨，而且有效运用了传单的写作惯用手法。题目要求写一份传单，文章的很多地方可以反映传单的文体特征，比如开头的称呼、以点列式交代活动内容、结尾呼吁读者参加活动、留下署名和日期等。因此，文章在形式部分可以获得满分5分。

Teacher's comment

Criterion A – Language: The complex sentences used indicate that the student has a good mastery of language. There are occasionally inaccurate phrases, such as "看的" should be "显得". Overall, the essay contains ample vocabulary, appropriate four-character idiomatic expressions and colloquial language, which show the student has a solid language foundation.

Criterion B – Content: A great article with clear structure conveying the right ideas, such as pointing out in the opening that the flyer aims at "鼓励学生参加活动", followed by the activity details and methods of participation. In general, the content is sufficient and conveys appropriate information to readers. Thus, the writing purpose is achieved.

Criterion C – Format: Overall, the text type can be identified easily, and is composed appropriately in the format of a flyer. The question requires the candidate to write a flyer and many areas reflect the text type stylistic features, such as the opening salutation, using point form to introduce the activity content, call for readers to join the activities at the end, signature and date. Therefore, this part scored 5 points.

中文B HL B部分

根据下面的陈述，谈谈你个人的看法，并且做出解释和论证。你可以自由选择你所学过的任何一种文本类型作答。字数为180-300字。

1. "人类为了开采石油、煤矿等资源，常常造成对自然环境的破坏。"

现时社会发展飞速，人类拥有了更先进的技术来开采资源。虽然开采这些资源对人类生活有帮助，但同时也对自然环境造成不少破坏。

> 入题直接，观点明确。此处连接词"虽然……但……"用得好。

首先，人们在开采资源的过程中造成了不少空气污染。比如中国大陆的空气污染令人触目惊心，人们在白天也伸手不见五指。而对自然环境破坏很大的酸雨的产生，也跟人类过度开采自然资源有关。其次，二氧化碳和其他污染物体更导致全球暖化的产生。它会影响农作物生产，更使得不少动植物濒临灭绝。另外，连番的石油漏油事故也造成严重的水污染。海面厚厚的石油使不少海洋生物窒息死亡，严重破坏了海洋生态平衡。

> 使用俗语，让语言增色不少。

总之，人类开采自然资源时产生了很多环境问题。因此，我们不能置之不理，袖手旁观。我们要以身作则，不浪费资源，呼吁社会关注此事，促使政府制定更多政策保护环境，大家携手去拯救地球！

> 连接词的使用，让文章的结构清晰。

评分标准 Benchmark*

评分标准 Benchmark	该项标准总分 Total sub-score	本文得分 Score in the student's essay
语言 Language	10	9
论证 Argument	10	9
总分 Total score	20	18

* 此评分标准只供参考，并不代表IBO官方立场。 *The scores are for reference only and do not in any way represent the views of IBO.*

教师点评

标准A语言：学生准确有效地使用了广泛的词汇，很少出错。文章词汇比较丰富，使用了一系列的成语和俗语，如"伸手不见五指"、"置之不理"、"袖手旁观"等。

标准B论证：文章层次清晰，开头入题直接，明确提出观点，紧接着第二段分析问题，结尾总结提出解决方案，显得层层深入。同时，学生善于使用一些连接词，比如"首先"、"其次"、"另外"、"总之"等，使得论证条理清晰连贯。另外，如果论证过程提及问题关键词"开采石油、煤矿"，会比较紧贴题目。

Teacher's comment

Criterion A – Language: The student uses a wide range of vocabulary and there are few mistakes. The article also contains various idioms and sayings, such as "伸手不见五指"，"置之不理"，"袖手旁观", and so on.

Criterion B – Argument: The article shows a clear progression. The student first introduces the topic in the opening, analyzes the problems in the following paragraph and offers solutions in the end. The student also made good use of transitions, such as "首先"，"其次"，"另外"，"总之", etc., which enhanced the flow of arguments. It would have been better if the keywords "开采石油、煤矿" were used.

2. "有人建议在中国的中小学应设置繁体字教育，目的是将中国文化的根传下去。"

"我认为"的使用，使得文章开头便明确提出观点。

　　随着简体字的使用日益增多，有些人开始担心这趋势会造成中国文化的一种隔断。我认为人们两种文字都应懂得写和看，因此我们应在中国大陆将繁体字列入中小学课程。

应加"除了"。

　　首先，学习繁体字有助人们理解中国文化，还对人们学中文非常有利。繁体字是中国文化的基础，明白到繁体字的由来，就能明白到中国文化的由来。繁体字可说是中国古代文化的缩影，因此，学习繁体字可让人们体会中国过去灿烂的文明。繁体字背后的由来更有助人们学习中文，它不像简体字一样每个字都要死记。其次，繁体字比简体字有美感，因此更适用于书法。

语法不规范，应改为"了解繁体字的由来就能更好地感受中国文化"。

连接词的正确使用，能有助于观点的清晰表达。

　　但另一方面，大部分50岁以下的大陆人也是使用简体字，所以为了方便交流，人们也应懂得写简体字。另外，简体写在书写较方便，有助外国人学习汉字。

搭配不正确，应为"简体字的书写较方便"。

　　总括而言，我认为人们应懂得通用简体字和繁体字，让我们丰富多彩的中国文化继续流传下去。

议论文的结尾通常用总结性的词语带出观点，比如"总括而言"、"总而言之"、"总之"等。

评分标准 Benchmark*

评分标准 Benchmark	该项标准总分 Total sub-score	本文得分 Score in the student's essay
语言 Language	10	8
论证 Argument	10	8
总分 Total score	20	16

* 此评分标准只供参考，并不代表IBO官方立场。 *The scores are for reference only and do not in any way represent the views of IBO.*

教师点评

标准A语言：学生比较有效地掌握了所学语言，虽然有一些不规范的词句，但整体语言通顺且所用词汇广泛。如果能多用一些四字成语，可以让语言上一个得分等级。

标准B论证：文章总体结构清晰，观点明确，论证有条理。思想观点得到了较好的论证，只是文章第二段的内容可更简洁，对简体字的论证内容较繁体字少，显得详略不当。

Teacher's comment

Criterion A – Language: The student shows good mastery of the language. The overall fluency is good and a broad range of vocabulary is used, although there are signs of improper sentence structures. With more four-character idiomatic expressions, the language score could be higher.

Criterion B – Argument: Overall, the article is well-structured. The viewpoints are clear and opinions are supported with good arguments. However, the second paragraph can be more concise. When talking about the use of traditional and simplified Chinese characters, more focus is placed on the former, which causes a slight disproportion in the content.

3. "现今社会学生经常要面对各种压力。"

压力是所有学生在求学时期中会面对的东西，可是，无论压力是来自家长、老师、同学或自己，我认为最重要的是学生们处理压力的能力和技巧。

面临考试季节，压力都对学生们有着一定的影响。若学生们不合适地处理压力，便会引智失眠、集中力下降及食欲不震等后果。要处理压力及避免压力引致负面影响，学生应先了解压力的主要来源。一项青少年调查发现大部分中学生压力来自自己及父母对成绩的要求，其次来自课程的要求及与同学的竞争。

我认为学生应从另一角度来看待成绩，不应把成绩当成个人成败的唯一衡量标准。此外，家长也不应给孩子带来不必要的压力，更不应以重罚的方式来强迫孩子拿取好成绩。这不但令学生过于受压，还会令学生忘记求学的最终目的是学到知识。除此之外，社会及学校应多给学生援助，多举辨多元化的活动及有助纾缓压力的节目。学生们也应多参与课外活动，如球类活动及学乐器等，避免考试压力对成绩甚至健康做成负面影响。

我认为学生们若想正确处理压力，要先了解压力的来源并对正下药，父母与学校也要给予协助，帮助学生减轻压力。

关联词失当，应改为"不管"。

此句可删去，显得更简洁。

错别字，应为"食欲不振"。

"来自"前面应加"主要"。

错别字，应为"举办"。

用词不恰当，应改为"造成"。

恰当的成语运用令人眼前一亮，可惜出现了错别字，应改为"对症下药"。

应为"还是"。

错别字，应为"引致"。

举例子，使得论证增强说服力。

语言不够准确，"从另一角度"改为"正确"。

评分标准 Benchmark*

评分标准 Benchmark	该项标准总分 Total sub-score	本文得分 Score in the student's essay
语言 Language	10	7
论证 Argument	10	7
总分 Total score	20	14

* 此评分标准只供参考，并不代表IBO官方立场。 *The scores are for reference only and do not in any way represent the views of IBO.*

教师点评

标准A语言：学生一定程度上掌握了所学语言，有时会出现语句不通顺和错别字的问题，比如用错"无论"，把"对症下药"写成"对正下药"。这些问题一定要改正，否则将影响语言的得分。另外，文章句式较简单，应该长短句结合，多使用复杂句结构。

标准B论证：文章论证大体清晰，运用举例子、讲道理的论证方法，论点比较可信。可惜有一些内容显得累赘，比如第二段开头一句应删去，使思想观点得以更好。

Teacher's comment

Criterion A – Language: To a certain extent, the student has mastered the language, although some sentences are not fluent and some words are wrongly written, such as the improper use of "无论", and "对症下药" written as "对正下药". If not rectified, these problems will affect the language score. In addition, the sentences are relatively simple. It would be better to incorporate more sentence structures and complex sentences.

Criterion B – Argument: The article's argumentation is generally clear, using examples, logical reasoning and convincing arguments. However, some parts are a bit long-winded and can be omitted, such as the beginning sentence in the second paragraph.

4. "现代年轻人过分依赖科技和互联网，以致于失去与人沟通的能力。"

现代，年轻人的生活绝对离不开科技和互联网。因为科技的发展非常迅速，现代年轻人不但只在家会用到电脑，在学校也可用到。

对我来说，科技和互联网的好处是它是一个沟通拿手的工具。以如说现代非常流行的面书可以帮我寻找和联络出了国或失散多年的好朋友。我觉得这种社交工具不但不会使年轻人失去与人沟通的能力，而可让他们的社会生活更加充实，和加强他们交流的能力。

除了沟通，科技也可为年轻人提供更多的道理去获取各种的资料。

一方面科技是一样有用的工具，另一方面对年轻人带来很太的影响，使他们失去与人沟通的能力。比如说，在互联网上有很多游戏给青少年玩。大部分游戏与现实相差很远，比如角色扮演游戏。这些游戏使人沉迷，甚至使人不想与别人沟通。所以，互联网也有它的坏处。

总的来说，我觉得互联网和科技不是一样不好的东茜，只要不要过分沉迷就可以了。

批注：

- 应删去"，"。
- 应为"不仅"。
- 用词不当，应改为"方便快捷的沟通工具"。
- 应为"比如"。
- 语法错误，"和"不能用在短句的开头，应改为"还可"。
- 关联词不当，应为"反而"。
- 用词不当，应改为"渠道"。
- 语言不通顺，应删去"的"。
- 词语错误，应为"大"。
- 此观点跟上文"不会使年轻人失去与人沟通的能力"相矛盾。
- 错别字，应为"东西"。

评分标准 Benchmark*

评分标准 Benchmark	该项标准总分 Total sub-score	本文得分 Score in the student's essay
语言 Language	10	6
论证 Argument	10	6
总分 Total score	20	12

*此评分标准只供参考，并不代表IBO官方立场。 *The scores are for reference only and do not in any way represent the views of IBO.*

教师点评

标准A语言： 文章出现错别字、语法错误和语句不通顺的情况较多，例如将"比如"写成"以如"、"大"写成"太"等。

标准B论证： 论证虽然有明显的条理安排，也会使用一些连接词来连接上下文，但是文章的观点存在前后矛盾的情况，比如第四段和第二段的论点相矛盾，导致文章观点不清晰。同时，文章的事例不足够，围绕论点的解释亦可更详尽。

Teacher's comment

Criterion A – Language: There are many wrongly written words and grammatical errors and the sentences lack fluency. "比如" is wrongly written as "以如"，"大" written as "太", etc.

Criterion B – Argument: Although the arguments are organized clearly and properly linked with conjunctions, some arguments are contradictory. For instance, the argument in the fourth paragraph contradicts with second paragraph, so the viewpoints are not clear. There are also not enough examples to support the arguments; more detailed explanation can be given.